Stop, Look, and Listen

This is Life in Kansas

Gus A. van der Hoeven

Stop, Look, and Listen
This is Life in Kansas

Black and White Edition
ISBN 978-1-7365419-2-0

van der Hoeven Publishing

Front cover sketch by Guo Liang Wu. Photo on back cover courtesy of Kansas State Collegian. All other photos courtesy of the author.

Dedication

To Anneke and all Kansans

There is no question that my "observation, seeing things" comes from my mother's side. The gift of "presentation," I received from my father. The result is this selection of vignettes--short stories--as I experienced them, saw them at that moment. This collection is to share with Kansans and people who visit the uniqueness of our state--Land of the Southwind. Don't just drive through it, but "Stop, Look, and Listen--this is our state, Kansas."

Silent hills near Ellsworth, Kansas.

Table of Contents

Foreword

When I became the radio agriculture director for Kansas State University Extension in 1983, I inherited a handful of tasks from my predecessor, Paul DeWeese, including the production of *Stop, Look, and Listen*...the genesis of which is entirely credited to Paul. The original premise of this program was for Paul and Gus to meet in the KSAC (later, KKSU) studio weekly to record a free-lance conversation about the unique features of rural Kansas. However, it quickly evolved into a colorful monologue built around Gus' uncanny sense of observation and perception. He harnessed his ability to see beyond the surface of things we think of as normal or routine, and to describe them in ways we would have never thought about. Add to that his masterful delivery of his writings, and the result was an immensely popular radio program with legions of followers. Even after his retirement from Kansas State University in 2000, Gus continued to share his talents with the radio audience for the next 20 years.

Gus is a Kansan...by choice. And is proud of it, as much so as he is of his Dutch heritage. The pathway here from the Netherlands, where he grew up the son of a Presbyterian minister, was hardly direct. As a young man, he ventured to Australia to work on a dairy, and to study at an agricultural college. Then, it was on to the United States, to sharpen his landscape horticulture skills via a five-year stint with a commercial nursery and through his studies at Virginia Tech University. In 1974, and with doctoral degree in hand, he and his wife, Anneke, packed up their young family and headed to the new adventures that awaited in Kansas.

Gus had been hired as the new Extension Landscape Specialist at K-State. During his 26-year career in that capacity, he was *the* source of choice for information and guidance on landscape design and layout for homes, farmsteads, and just about any other venue. The results of his work are showcased statewide, from the most modest of rural dwellings to Cedar Crest, the mansion home of the Kansas governor. As one whose own home landscape was expertly designed by Gus to perfectly complement our house and the surrounding Flint Hills, I can be counted among the scores of satisfied customers.

It takes an uncommonly creative mind to come up with an insightful and entertaining radio essay each and every week. But for 45 years, Gus has done precisely that. The selections included in this book are just as illustrative in print as they were over the radio. They offer a window to life in rural Kansas that could only come from someone with the deepest appreciation for the sights and sounds of the state, and for its people.

Eric Atkinson is now in his 38th year as Extension radio specialist at Kansas State University, and producer and host of *Agriculture Today*.

Well built. Well maintained.

Preface

It was April 1, 1974, when I moved to Kansas and started to work for Kansas State University. Actually, the way I saw it, I started to work for and with the people of Kansas as the Extension Specialist in Landscape and Environmental Horticulture. I worked with the county Extension personnel and horticulture industry as a catalyst for landscape and environmental horticulture.

What struck me from the beginning of my Extension travels and contacts was hearing people apologizing for the state of Kansas. Those were *Kansans* apologizing! At first, I did not get it, but when I asked, they would say, "Oh, our state is not as exciting or beautiful as others." I assumed they thought because I had a Dutch accent that I expected lots of flowers and green grass everywhere. I did not. My years of living in Australia and reading up on Kansas before I accepted my position had prepared me well.

My travels through the state from corner to corner gave me the opportunity to study the different physiographic regions. As I traveled, I used my 35mm camera to record things I saw. I only wished I had taken more photos. But those I did not take I hold in my head.

I talked with Paul DeWeese, with whom I had started to do educational radio programs on the then radio station, K.S.A.C. I told him, "I want to start a brief, weekly program to show people how unique Kansas is. How I see it having just arrived here."

Out of our conversation came the weekly program, *Stop, Look, and Listen--this is our state, Kansas,* which you can now listen to on *Agriculture Today,* aired with and introduced by my friend, Eric Atkinson. My critic, as well as my friend, Christy Dipman, types the weekly vignettes.

In the form of this book, I want to share some of them. Maybe as you read an old one, you will recognize or remember hearing it--the accent and the place. When you read them, don't read too many at once. Read one or two--no more. Take note of the dates they aired; and remember, you are reading the spoken word. These were radio programs.

When my daughter, Talina, sorted, selected, and edited the vignettes for this book, she said, "Dad, it amazes me how often in your stories you made a connection to the old country--the Netherlands, where you grew up."

It is true. But remember, I came to this country. I came to Kansas. It was not for nothing that after I had arrived in Kansas and after meeting people, several times they remarked, "You sound just like my granddad." Kansas is still so young.

When I called home in the evening after my first long day driving to Kansas to begin working, I learned that there had been storms with tornado activity near Blacksburg, Virginia. That was where I had started my drive from early that morning and where Anneke and the kids still were, where home was. Whatever weather Kansas has thrown at me these past 46 years, I have always remembered that call home. Our weather is unique, it gives our state its name, "Land of the South Wind," Kansas.

The fact that 98 percent of Kansas land is in private ownership is, I think, one of the unique aspects of our state. It places the responsibility for its landscape on the owners, the people. And being an agricultural state, its land is cultivated, or where best suited, used for grazing as on the Flint Hills, requiring range management including burning. Those who have burned know the hazy smoke as seen through a setting sun.

Through telling stories and sharing photographs I hope to show what our state is about. It is true each of us may see it differently but remember, our state, Kansas, is beautiful. Maybe you have to look a bit harder but once you see it, or hear it, or catch the scent of ripe wheat, you will never forget it.

Kansas

Big Blue flowing into the Kansas River.

April 1, 1992

Our State, Kansas--and that is No Joke

April 1st is a date I like to celebrate. Not so much because of the pranks and jokes, but on April 1, 1974, I came to Kansas and joined the Horticulture Department at Kansas State University as the Extension Specialist in Landscape and Environmental Horticulture. That was years ago. It is a strange thought that in all my life I have never lived this long in one place and in the same home. Until now, never having seen trees and shrubs I planted grow full-size. Each time I moved on, leaving all that I had begun behind.

I think it was a year ago, during a discussion of perhaps moving further out of town and building on the farm, that our youngest son made the remark, "You are not thinking about moving, are you? I don't want to go. I have always lived in this home. I would not know my way around in the dark in another house." Lately, he has come around, but that is because, with his continued growing interest in horses, he sees the possibilities of moving to the farm.

But we are not packing yet and as I walk around this house, I observe all the improvements we have made such as the arbor, the big flower box which cheers up the patio in summer, the patio doors, the deck. Stopping by the shade trees we planted years ago, these now give shade. The cedars and pine along the north-northwest stop the cold winds. And all the shrubs and flowering trees we planted have created a bird sanctuary and our feeders provide a welfare system that even the squirrels enjoy.

In the summer after a hot day, we sit on our deck and watch the moon sail by us, looking down over the town of Manhattan. Its light spreads out to the west now where years ago there was only a wheat field and beyond it hills. Our hill, where we live, gives us a beautiful view over the valley.

The day we drove up with the U-haul truck and trailer hauling furniture, kids, dogs, and a pony, the house stood bare on the hill. To landscape here was a challenge: to plan, dig, plant, and water. A tree I am especially fond of is a Bur oak planted as an acorn. It is now 30 feet tall. The tree has already produced small crops of acorns, some of which seeded themselves and have grown into seedlings. I must dig one up and plant it on the farm just in case we move there someday. An offspring of the Bur oak I planted as a seed years ago on our hill.

I don't know where these many years have gone, but they have been good. Working throughout the state, I now drive familiar roads. But every now and then I hit a road never traveled before and it thrills me. I see a valley, hill, or flat fields; further noticing a barn or clump of trees, then farmers on their tractors working their fields. At night, passing the lights of a town, it is all familiar. It is a bit like my son said with our house--not being able to find his way in the dark if we moved. After all these years, I know the roads and am familiar with the geography of the state. Feeling comfortable following an unfamiliar road because I know where it ultimately will go. Even driving gravel roads, sooner or later you'll reach a hard-surface road.

Last week, after working in Elkhart, southwest Kansas, I drove through the Cimarron National Grassland and followed the gravel roads going east. Intersecting with a hard-surface road, I then went north to Ulysses. You just can't miss it. You can't get lost. It is beautiful country, a beautiful state. Our state, Kansas, and that is no joke!

February 7, 1990

"Pops, YOU are the Cat's Meow in Holton"

Whenever people work together, there is good-natured ribbing or joking. It is the oil that keeps the machinery of human interaction running smoothly. It releases tension, helps overcome anxieties, and keeps things in perspective so you don't take yourself too seriously.

I receive plenty of ribbing from my colleagues and admit to "dishing it out," perhaps more than one should. Occasionally, I get ribbed about this radio program, *Stop, Look, and Listen*, which has been aired every Wednesday afternoon for nearly 45 years. They like to say, "van der Hoeven, no one listens to you," or they ask, "Who does listen to you?"

The interesting thing is I never have worried about it. Yes, one thinks about it, but I don't worry about who may listen to any of my programs. It is like painting a scene. You paint because you feel the urge to paint the object, the view, the light fall, the moment, the happening, or the incident. When you are finished and the paint is dry, you sell it or give it away.

In the same spirit, I put my observations on the airwaves and the words are free to listen to. My standard remark when I meet someone who says, "I heard you on the radio, etc.," is, "Well, remember you can always turn the knob and switch me off!"

It is kind of a silly answer. Of course, I'm glad when someone says, "I like that program," following with a short discussion on the radio vignette heard.

When asked, "What is your program going to be on this week?" I never tell. It would lose some of the magic. Once in a while, I may suggest someone listen to a program, feeling that person would get pleasure from it.

But you never know how people will respond. A few years ago, after a heavy and beautiful snowfall, I described the white, quiet Kansas winter scene. Sharing how I looked at large snowflakes falling on the sleeves of my jacket. Going on to describe the structure of the different snowflakes as I saw them slowly touch down, touch down. Those were large, perfect flakes and they were beautiful to look at.

Sometime later, a colleague remarked about that program and said, "How, how could you talk about a snowflake?" Perhaps he meant there were more important things to talk about. Maybe so, but I liked to talk about snowflakes that day because I had seen them and I happen to believe it is sometimes good for all of us to look at simple, yet complicated, snowflakes.

I never know who listens, but if you do, you should know there are two of us who put this program together. Eric Atkinson leads me into the program, and he is the voice you hear at the end. Our program is a cooperative effort. Yes, I'll take the blame if you turn the knob and search for another station. Of course, that is what I tell my colleagues when they rib me about the program and ask who listens. But I have news for them. I know I have an audience and I know who they are.

My daughter-in-law gets milk direct from a dairy. She knows the way through the milking shed and gets the milk out of the cooler. Not so long ago she went to get milk in the afternoon. Upon entering the dairy parlor, she found it empty. No one was around, only the radio; KKSU on the 580 dial was on. As she entered the clean, empty milk parlor, she heard my voice. Everything was set up for the evening's milking, but there was no one there. No one except a number of cats perched on the milking stools. One cat on each stool, they all had their heads cocked sideways and they were listening. Listening to me!

"Pops, you are the cat's meow in Holton."

January 30, 2002

West-more-land

The country road wound over the hills, dipping every now and then to cross a low running stream. I was in no hurry but people who were, passed me. The radio announcer confirmed it was a beautiful day with temperatures in the high 60s in Manhattan. I had rolled the truck window down to feel the cool air go by.

It was a perfect day, and I was on my way to Westmoreland. Being used to saying Westmoreland, I forget what I am saying: West, West-more-land. Although the name seems to be very clear as far as its meaning goes, it is the name of a county in England.

A John McKimmas, who had been postmaster for Westmoreland County in Pennsylvania, gave the name to the county seat in Pottawatomie County. I wonder if when he gave the name, he felt homesick for the county with the same name in Pennsylvania. Or did he say, "West-more-land," as they must have said in Pennsylvania? "West, West-more-land" with other words, "Go west young men."

Our farm is in Pottawatomie County so, once in a while, I do have to go to Westmoreland to sign papers or to talk to someone in the agriculture office. I also simply like the drive as it is a beautiful road if you drive Flush Road past the church and cemetery.

The day I drove down, the church stood out against the blue sky, the steeple pointing to the heavens. From the distance, I saw all the cars parked and as I got closer, the few cars on the road were momentarily stopped as people filed out of the church and crossed the road to enter the cemetery. There was a funeral on this beautiful day. I remembered reading of a tragedy and wondered if this was the funeral or memorial service as people thronged around the fresh grave.

As a passerby, I sensed the feeling of community. The wide-open country with the farm homes, the cemetery as the gathering place, the tall church framed in by a few trees as the heart with its steeple pointing up--high. When the road was clear, I continued my drive and slowly passed by, remembering the annual community meal of chicken and all the trimmings I had enjoyed together with friends years ago. Across the road, the people stood together--a cemetery is a gathering place. It brings us together and it will harvest or gather us in time. As I drove on, the landscape had not changed, but my feelings had changed. My thoughts were still with the people of the community I had just passed, the people in black.

After finishing in Westmoreland, I drove home by going west, returning to Manhattan by Route 13. A road I also know well, driving it often. Our farm lies west of Route 13, west over the hills. There is a lot more land west. Actually, if you look at a map, there is a lot of land and not too many roads. That is what makes that part of the county very interesting. I believe at one time, the land, the hills, were thought about as a possible prairie or tall grass park. West-more-land.

I wonder what the county in England and the county in Pennsylvania look like with the same name, and if there was West-more-land. We know it was the case for the town in Pennsylvania. They came a long way, but indeed there was more land out west. Of course, the same was true for the town in England, except that after they reached England's west coast, they had to cross the ocean until they reached the New World and then from the eastern shores, indeed, people found West-more-land.

Remember, when Lewis and Clark made their journey. They came back a year later finding the frontier from which they had set out had in that one year moved several hundred miles further west. West-more-land.

I don't think I can drive to or through Westmoreland without thinking about the town's name and not turn my back to the east but look west. For me, the town will always be West-more-land. Go and look at it for yourself, and when you drive west on the Westmoreland road, you will pass smiling faces on tree stumps--smile!

September 26, 1990

A Kansas Day

"The wind must have blown down from Baba's farm while we were sleeping."

Is it right to teach little children to love a certain place? Why do people love a country, a state, a province more than another? Why do people get homesick? What do they get homesick for? Is it the people, family, friends they once knew? It seems that land does not change as much, which of course is not true because the moment you leave for an extended separation, things start to change.

You can never go home again. You can go back to the mountains. You can go back to the seas. You can go back to the forest. You can go back to the plains. Yes, you can even go back to Kansas, although here also things change. Sometimes faster than I care to acknowledge or accept.

Last Monday, I noticed those beautiful healthy campus trees marked with huge red crosses. They were marked to be bulldozed out. By noon, the first one was dozed over and the chainsaws were tearing into it, screaming noise. All for the sake of another parking lot extension. It makes me sick.

I called the campus planning office and asked why the trees were cut down. Parking lot expansion and reorganization of a drainage ditch, which on paper is called a natural stream, was the answer. I also learned that they were planning to landscape the new site and it would all look beautiful again. Someday!

You know, I hate to see healthy, beautiful, old trees cut down. So much for change. You see, perhaps it is much better not to love someplace. Then it does not hurt when places are destroyed or are modified in the name of progress.

So why have we, with their parent's consent, instilled a growing loyalty to Kansas in our grandchildren who live out-of-state in Ohio? Have these children heard their parents talk of the joy of living in Kansas? Do the children associate the love of their grandparents who live in Kansas with loving Kansas? Or is it really something else? Are these children more sensitive about the Kansas they know?

My daughter and her husband, both Kansans, live near Columbus, Ohio. Recently, she wrote a letter and described the following scene. I quote, "Yesterday morning was a beautiful day--true blue sky, small clouds scattered around and a different feel and look to the air and sky." And Joseph, our grandson, stops on the porch step, breathes in really deep, and turns to his mother with a smile to say, "It's a Kansas day, Mommie, the wind must have blown from Baba's farm while we were sleeping." My daughter finishes her note with, "Needless to say, we all had a good day after that profound beginning."

You know, when it is raining and just miserable weather in Ohio and there is plenty of it, Joseph will say, "This is not a Kansas day. Why don't we go back to Kansas?" He is only five years old. The last three years of which he lived in Ohio.

Besides trips back, you wonder what gives him this strong affinity--love--for our state. If it is in the genes, then his father's genes must be dominant because my daughter has some affinity to the old country. But yes, she too loves Kansas and I suspect pours it into her children with breakfast cornflakes and home-baked bread, all made from Kansas wheat, ripened in Kansas wind, and Kansas sunshine. Yes, it may sadden you when you must move on, but it is better to have loved a place than not to have known it at all.

Okay, Joseph. If you need more Kansas days, I'll do what I can. This evening I'll climb the hill on the farm and I'll blow and blow and blow and when you wake up tomorrow morning, there will be another Kansas day with sunshine blown to you all the way from Baba's farm.

February 6, 2002

Cowboy

He was a short, stocky man wearing a warm flannel shirt, faded old jeans with scuffed, knee-high cowboy boots. I nearly walked into him as he purposefully walked the sidewalk, knowing where he wanted to be. He looked like a cowboy from years ago, one who would have been painted by Charles Russell, the western artist.

There was no question he was for real, with his tinkling spurs. Not asking, I did wonder, though, where his horse was. There was no time to ask as he pushed by me and asking would have been a bad joke because the man was, even though old, a cowboy.

It is now several weeks ago that I passed him on the sidewalk in a small Kansas town where I was strolling down Main Street, not in any hurry.

The man seemed to come from nowhere as he moved quickly past me, slightly bow-legged, wearing tight jeans and tall boots. The tall, knee-high boots give the protection and comfort you want when you ride. Tall boots keep your jeans in and prevent the stirrup leather from chaffing your legs. My son, who rides show jumpers, wears tall boots--very tall boots--as he has very long legs. The only difference is his are English style riding boots, but the need is the same protection. Yes, the riders look good with their britches, the black fitted jacket, and the hard hat when they ride and jump in the show ring. Both men, cowboy and showjumper, are serious riders. Both men know horses, different horses indeed, but real horses.

Years ago, at a horse clinic in Weber Arena on campus, my son was asked to demonstrate English riding to the college cowboys who rode their quarter horses western style. They used the heavy western saddles and when they saw Martien's light English jumper saddle on his thoroughbred, a very tall horse, the cowboys looked up at him and asked, "How do you stay on?" Not only how do you stay on trotting and cantering, but galloping and when you jump?

When they realized that jumps were as high as five feet and over, they really looked puzzled. But it is all in the legs, the seat, and the hands. The saying "Legs like iron and hands like butter" is clearly shown in the picture on my desk where Martien and his horse, not even a big horse, clear a jump. Both horse and rider are very concentrated, but you can see the very gentle yet firm and controlled contact with the horse's mouth. It is a beautiful photo of a beautiful moment, the taking of a high jump by horse and man. The instant moment of sailing through the air, clearing the poles, to come down and gallop to the next jump--over again.

But just as real as Martien is, so was the cowboy. The old man with many years of riding, rounding up, sorting, pushing, separating cattle. I wonder if he still had all his fingers to prove that he knew what he was doing when roping or, if by chance that one time, he perhaps did get his finger crushed as he wrapped the rope smartly around the horn of his western saddle and got caught as his horse and the cow pulled the rope tight between them.

The town where I walked down Main Street was not Dodge City, but a town in the cattle country of Chase County, in the Flint Hills. Whenever I have to go south, I prefer to take the scenic Route 177 through Council Grove, Strong City, Cottonwood Falls, Matfield Green, and Cassoday.

While in Australia last fall, visiting old familiar places, I walked into what we would call a western store. My interest was the Australian saddle. I walked the store and looked at all the tack, noticing and remembering the similarities and differences, knowing darn well that my wife would really question my sanity if I bought yet another saddle. But I could not leave the store without browsing through the books displayed on a table. The book, *Big Mobs, The Story of Australian Cattlemen*, caught my attention. The book is written by Glen McLaren, an Australian, but it has comments by Professor Jim Hoy from Emporia State University, in Kansas. Apparently, the two men knew each other and had a shared interest in cowboys, or stockmen as they are called down under.

There are differences between the stockman and the cowboy, but they have much in common. They both worked cattle in wide-open country--in Australia, longer than here--where fences intruded on the Plains. They were and still are superb horsemen and they dressed for work, which was the impression the stocky, wiry man I met on the sidewalk gave me with his tall boots and tinkling spurs. He was going somewhere!

December 12, 2012

Kansas Wonders

Some time ago, I listened in on a program stressing the eight wonders of Kansas. Earlier, I had heard that people could share their particular Kansas feature and send it to the selection committee. The eight which were ultimately selected would then become the "Eight Wonders of Kansas."

The program, which I just happened to catch on TV, caused me to sit down and think. What are my Kansas wonders? I very quickly came up with four, but then I slowed down coming up with more. I slept a bit and forgot them by morning. Very amazing. But as the day wore on, they came back to me. Right now, I am past the eight, which doesn't bother me, and still coming up with more. Mind you, a few of them might be claimed by other states as well. No problem, I am very willing to share a wonder. I am not asking you to agree with me, remember these are my wonders! Here they are in no particular order.

My first wonder is growing wheat. Here we have this plump wheat seed kernel. It goes into the ground with care and then germinates under the right conditions. And then there come the first shoots just peeking above the dark earth. Millions and millions of kernels all do the same at about the same time. What was first a dark, dormant field suddenly is a green blanket, growing, growing. Oh, yes! The kernel does need moisture, but it is oh, so willing to hold out for hope. Then when the crop is ready, the one kernel gives an ample return. It is a wonder!

Our sunsets and sunrises are a wonder to behold in the Kansas sky. They can be so colorful and dramatic that one holds one's breath. As you look, the sunset or sunrise will change through the most fantastic color kaleidoscope right before your eyes.

Another big wonder is, in the spring, how one tiny match can set the prairie pasture on fire. It turns the hills black. Then, in a few weeks, the hills turn soft green and by mid-summer, those hills are covered again with tall waving grasses. That is a wonder.

Then there is the moon, which on a cold or warm night encourages its own night symphony of sounds from the yipping coyote or the hooting owl. If you are lucky enough to stand next to a calm horse, you can see the moonlight reflect in the horse's quiet eyes.

Another wonder is the silence. In Kansas, you can get away from it all and listen and hear nothing. Oh, you do hear something, but it is not noise. It is silence. Very, very peaceful.

We have our limestone buildings. The beautiful houses and the imposing courthouses with their specific architectural features stand in Kansas. Just as important are the barns, bridges, and fences of the Post Rock Country. If you look, you can see and study the fossils and wonder, wonder that once much of Kansas was part of the big shallow inland sea. We see it in the limestone strata wherever there is a road cut through the rocky limestone. Often, we rush by, but I like to see the geology students climbing along the road, chipping, and cutting small pieces, and wonder what they have found. A piece of rock laid down millions of years ago as sediment. They study and wonder.

Thanks to the historic movie by Ken Burns about the Dust Bowl, I wonder how the era of blowing soil was stopped and that the land healed. Too much soil had blown away; restoring was not possible. The land would never be again what it had been for thousands of years, but through the hard work of men and women, they were able to heal what had been ripped open. The land became viable land again. That too is a wonder.

To me, the Kansas roads are a wonder. If you think about it, roads connect us. Roads carry ideas or at least they did more before we had computers. Roads give us horizons and vistas and take us from here to there. I remember years ago, an old Dutch farmer friend asking me after I had lived in Australia and then in Virginia,

"Gus, how do you find your way everywhere?" I simply follow the roads.

I think the people of Kansas are a wonder. I know farmers who can take any piece of farm machinery apart, fix what is wrong, and then joyfully put them together again without holding an extra bolt or nut. They are inventive. They take care of the land and the livestock. They have a lot of knowledge and common sense is passed on. But there is something in their genes, they are Kansas farmers who hang in there with hope, determination, and knowledge.

I think the Kansas trails are a wonder. There is the Oregon Trail, the Santa Fe, and all the different cattle trails leading into cattle towns. The history and the long tales about the cowboys are a wonder. Those tales are the stories, not the tails of steers and cows. We can celebrate our history in a way that it enriches us. Share it with other people so they too might have an insight into what Kansas is all about.

September 25, 1985

Douglas Firs

The epitaph on the burial stone reads, "The greatest pleasure in life is to do a good deed in secret and have it discovered by accident," and underneath it, "A citizen of two worlds." I stumbled across it by pure accident, in so far as one can say being inquisitive or interested is accidental.

Early in the morning, I had driven with the old nurseryman out of Abilene to Milford Lake. We drove Route 18. It was a spring morning, still somewhat misty but the day promised to be a pleasant Kansas day. Driving along, my companion, the nurseryman, said, "Somewhere here just north of the road lies an old, well-kept cemetery and it has beautiful large specimen Douglas fir trees standing in it." He continued, "I have never found out why and who planted them. But whoever did, must have loved trees." He pointed the cemetery out to me in the distance as we drove along. It was then that I decided on my way back I would stop and take a look.

Having finished my program and meeting, and having taken the old gentleman back, I retraced my tracks back home. I drove Route 18 again, going up and down with the hills and somehow, in the process, I must have picked up speed because suddenly, as I rose over another hill, I met the blue-gray car of the state trooper. A quick glance on my speedometer told me the reason why his red light was flashing. I had not looked at my speedometer. Instead, I was viewing the spring countryside and in the process, my foot must have become heavier and that, with the terrain of the up and down of the road, must have been the reason that my driving speed had gone way above the limit.

So, when I saw the flashing lights, I just relaxed and said to myself, "You got me, trooper; no excuses." But just as suddenly as I saw his flashing light dash over a ridge, I saw it disappear behind me and even though I slowed down to pull off the road and respond to his friendly request, "May I please see your driver's license?", I never saw him. He never turned around and pulled up behind me. I could not believe it. But he never came, and I slowly picked up speed again, up to the limit--and this time only up to it--and continued on my way.

A little further on, I saw the tall evergreen trees growing in the cemetery. I pulled off the road and followed the gravel road leading up to the slight slope toward the quiet place. Turning the engine off, there was only stillness around me with the tall trees--the Douglas firs--in front of me. I walked up to the trees and as my friend had done, wondered who had planted them, at a time when nurseries were few and far between. Whoever planted them must have wanted to make a statement about this place, the person buried, or the past. Standing there I could hear the spring breeze play through the needles, softly whistling.

The Douglas fir is native to the Rocky Mountains. It is a noble forest tree which in its native habitat can grow up to 200 feet in height. It likes well-drained, moist soils, and does not like windy sites. Knowing this, it looked like everything was wrong with the trees planted on a hill exposed to wind, except for the soil. Because I assumed that no one in his right mind would start a cemetery on a rocky hill.

As I walked through the cemetery, I kept looking at the tall, dark trees. They must have come from the Rockies, but who planted them and who cared for them initially? There was no answer, but it showed me once more which trees can grow in Kansas.

As I walked away, I passed the stone with the thought-provoking epitaph, which is also intriguing. It said, and I repeat, "The greatest pleasure in life is to do a good deed in secret and have it discovered by accident." "A citizen of two worlds." Which two worlds? The old world--the new--Europe--America? Or what? The mountains and plains? I have no answers, but that place, the small country cemetery has been on my mind the whole summer. It's just a small place in Kansas along Route 18.

May 16, 2007

Greensburg

Anything I write or say about Greensburg, its people, and the devastation by the tornado which flattened the town, has been said already by others on TV, radio, and newspapers. My thoughts go out to this community and the people. When a disaster strikes and one cannot run to neighbors because they are devastated too, it is very hard to take. It is tough.

My very, very small, insignificant inconvenience of this storm was groundwater seeping into the basement rooms and spending the night awake, watching TV and using the shop vac every fifteen minutes to suck up the water which came bubbling up between the basement tiles. It was a small price to pay to make sure that the water would not do any major damage.

I watched the storm's progress as shown on TV and I could, more or less, predict when it would end; when the red, yellow, green zone would have passed Manhattan. As the storm moved slowly northeast, Manhattan finally came clear in the early morning. It took until about 10:00 a.m. for the groundwater to stop seeping in through the hidden cracks of the concrete floor.

One might wonder how it is possible that water enters the basement when you live near the top of a hill. It is simply that the ground was saturated with the five inches plus of rain and the house sits on the hill like a cracked cup pushed into a dish of water. There is no problem with normal weather, but a storm like this flips the coin. Anyway, as I said, a very, very small inconvenience anyone in Greensburg would gladly have accepted for keeping their house.

It was not only houses people lost. They lost their homes, neighborhood, community, schools, places of work, and places of worship. As the TV reports kept showing the devastation of this town, there was one recurring scene that caught my eye. It was the aerial photo of the town showing the rubble on both sides of the streets. It was a panoramic shot. It reminded me of the old, browned photographs one sometimes can see hanging in the courthouse of an earlier town plat with a few houses built, but most land still vacant, waiting to be owned and built on. There was perspective in the scene of three or four streets on the square, fading toward the distance, everything gone but the elevators.

The Manhattan Mercury had a good article on the city administrator, Steve Hewitt, a young professional. Apparently, he has only lived in Greensburg for a year or so. But an experience like this will send deep roots into a community. It is like plowed ground being prepared for a new crop. The trouble is too much Round-up was used. I am not trying to be funny, but it would have been easier if a whole new town would be built on a clean plat. Now the rubbish has to be sifted through to salvage "home" and then it all needs to be cleaned up.

The only thing there is, or rather was, is some infrastructure and if there were aspects in the old town which needed to be addressed, now is the time to solve them. It will be a gigantic task. But people can rebuild a community.

In 1940, when Germany declared war on the Netherlands, the German air force bombed the harbor and center of Rotterdam. It was devastated, harbors and all. However, the fact that the harbors were destroyed, allowed for a major rebuilding effort starting in 1945, after the end of the war. Those new harbors and new city-center became the most modern of Western Europe which gave tremendous economic stimulation to Rotterdam. Remember that destruction was an act of war.

One does not ask for these things to happen, as in the case of Greensburg, as lives were lost. However, if guided and steered right, in years to come one may drive Route 54 and pass through a new Greensburg with wide streets and young, strong trees growing up together along the streets and in the city park. People who don't know the history of this town may wonder how come all the trees seem to be of similar age, height, and trunk diameter. I am sure somewhere there will be a sign to read to let people know what happened to the town and how the people rebuilt it; house by house, business by business, tree by tree, rebuilding a community with schools and churches.

In the center of the once destroyed part of Rotterdam stands a statue of a man stretching his arms into the sky from where the destruction came. It is a terrifying piece of sculpture--titled, *Monument to a Destroyed City*, created by Ossip Zadkine. It shows a surrealistic granite and bronze creature with a jagged hole where its heart was torn out, a wide-open shrieking mouth, and huge arms raised in despair as if trying to ward off the rain of death from the sky. The large statue has no heart, signifying the destruction of the heart of Rotterdam.

I am not comparing Greensburg to Rotterdam except that we know one can rebuild and replant trees. Plant trees--hardy, adapted trees--creating a beautiful park to go to and remember and leave rejuvenated, energized. Close to the very harbor of today's Rotterdam is a beautiful park.

I like that idea: trees, a park, a future for Greensburg! "Greens" burg!

December 16, 1987

Charles Rogers

If the walls of the old cowboy hotel in Ellsworth could talk, I wonder what stories they would tell.

The first time I stepped into the old hotel lobby, I did not really know what I was getting into. Curiously, I looked around until, from somewhere in the back, Charles came. Charles Rogers is the proprietor of the old Tucker Hotel in Ellsworth. Here Charles Rogers maintains his studio, gallery, and museum. After the introduction, he invited me to look around and I slowly wandered from room to room, gazing at the paintings and pencil drawings he had done.

Over the years, I made a few more stops and have bought a few reproductions of his drawings. I have the portrait of the old cowboy. I liked the face; open, honest, rugged, lined, and weather-beaten with warm, happy eyes, showing a deep warmth.

I bought the one of the hawk taking off or coming to rest on the limestone corner post. It is a scene seen so many times that I should be able to say if the bird is landing or taking off. I think what I like is the still action of the wings caught in motion but frozen on paper. I expect the hawk to flap once more and drift on out of the picture.

The one of the old cottonwood seen through a snowstorm is one I also like. You can feel the snow blowing around your head as you step out from behind the lee of the tree. The movement is there, even though again, it has been stilled on paper.

I have never framed the etchings bought, but I have looked at them every now and then, always glad to have them. In remembering my few talks with Rogers, he once told me, "We both seem to be independent people." I never knew how independent he was, but guessed he must have been quite independent, especially when young.

I think I will now frame the sketches I bought and hang them on the wall. I'll have to make room for them, but I think the time has come because Charles Rogers died last week at the age of 76. *The Manhattan Mercury* had an article on him, "Western Artist Rogers Dead at 76." It struck me because I passed through Ellsworth last week and I nearly stopped by to say hello and look at his paintings, but I pressed on my way to Garden City, the town in the wide-open country for which Charles returned from California when he got "horizon hungry." That is how he said it himself. When he told people that he was going back to Kansas, they said, "Kansas?" But he thought he had done some of his best work here. And nowhere else can you find the vistas you can get in Kansas and after being in California all those years, he got "horizon hungry."

As I sit in my room and look around me, I am surrounded by reproductions of paintings, etches, and photos all made by different people. It is what they saw and put on canvas or paper. When looking at my office walls, I look through their eyes. From Remington, a reproduction of a Native American scout on his pony gazing across a moonlit snowfield. Van Gogh, a reproduction sketch showing pollarded trees with small puddles of water. Another is a panoramic view of an old Swiss river town, Stein. An Australian scene of the outback--a stockman on horseback passing the eucalyptus trees, riding into the sun painted by Hans Heysen. Another the image of a joyous wedding scene by Pieter Bruegel. A black and white enlarged photo my daughter took of a very familiar view near the old home place in the Netherlands. All are scenes, places, captured by the eyes of the different people, artists. Some of the people are long gone, dead, but we still treasure their work. We still enjoy what they saw. When we look, it is as if we stand behind them.

Whenever I look at the sketches drawn by Charles Rogers, I look over his shoulder. It was his eyes which saw the snow, the hawk, the cowboy's face. He saw it for the moment and recognized it for what it was so that we could look at it again and again. Thank you, Charles Rogers, for coming back to Kansas and showing us on paper and canvas how you saw Kansas.

July 10, 2002

Water

I read that in southwest Kansas it was so dry that cracks were wide and deep enough in the soil that a man could easily stick his entire arm down into the crack and not feel the bottom. I wonder if a man could get away with singing the old song, "Cool, deep water," without getting thrown in jail for disturbing the peace, down Main Street in the southwest.

For a short period, we had some water restrictions here in Manhattan. But it was based on a technical breakdown and once repaired, people could go on watering their bleached and thirsty lawns and droopy shrubs. Nature is thirsty, except for the hardy who know what it takes to survive, even feel at home, in Kansas when it is hot.

On the farm, the spring has dried up all but a small murky pool where the deer come to drink and Kansa, the dog, likes to wet himself; where the frogs have retreated and where the blue heron strikes down for fast food. It is like a drive-through at McDonalds. You see the need for water on the farm by the folded leaves of prairie grasses, the wilting corn. We could use a good steady rain.

Hot as it is, I make sure when I go to the farm to take a can filled with cool water with me. I don't like to drink it cold, icy cold that is. Just cool will do for me. However, it seems that the grandchildren want it cold, icy cold, and not only that, they prefer canned drinks over plain water. As a matter of fact, there is a family saying originating from a nephew, who now is a father of children, that when he was young and complained he was thirsty, he was told by his father to go and drink water. This nephew responded with, "Drink water? Water is to wash your hands with." That happened a good 40 years ago and when I look around me, that attitude toward water seems to have grown exponentially.

Where I like to drink a cool glass of good tap water, my grandchildren reach automatically for a cold drink; Coke, Pepsi, or

Sprite, in the refrigerator, pull the lid--pppppsssss--and drink, take a sip. It seems that water is no longer associated with thirst. Not only that, it has to be cold, real cold.

With the grandchildren being thirsty, we have several large cartons of cold drinks stored in the garage from where they take and keep the fridge filled as one would pass the ammunition to a piece of artillery! I have studied this process with some amazement and near disbelief. I know when our own children were small, so-called "soft drinks" were very special treats, something to drink on a birthday or other special occasion.

As they say, the proof is in the pudding. I remember many years ago my youngest son, then seven or eight years old, was encouraged to jump his pony, Poptart, over some low rails with the promise of a Coke. That tempted him so much that instantly he overcame his reservations, turned his pony, cantered, and neatly cleared the low jump. Now, at 25 years of age, he trains and jumps Hunters and will jump a five to six-foot fence. But then, it was the specialty of the Coke which started him.

Another story about water is that of his eleven-year-old brother who took care of Star, one of our ponies, now long gone. It died of old age on our hill. One night when I checked the pony, I noticed that its drinking water in the bucket was not clean. We were about to have the family dinner and I quietly went inside, took a glass out of the cabinet, and filled it with the pony's dirty water. I placed that filled glass beside my son's plate while the other children had a glass of clean, sparkling water. I quietly told him that it was his water if he so chose to drink it, which of course he did not. It made an impression on all the children, but especially on him. The pony never had dirty water again.

Of course, ponies and horses don't drink Pepsi or Coke. Maybe they'll slobber a beer. But generally, you don't give canned drinks to them. But ever since the grandchildren are with us, I find empty cans standing on shelves.

I am now teaching them that recycling aluminum cans is a good thing. That aluminum is a resource that is mined. They do take it to heart, but I am surprised that they are not more concerned. Maybe they have not drunk enough good, clean water. Maybe we have to start all over and assure them that it is safe to drink water, that water is used for washing hands as well as drinking.

I told you earlier about the dried-up spring at the farm. It is just a small pool now, stagnant. However, I wonder if I cleared off the hundreds of trees and shrubs which grow above the rocky outcrop underneath which the spring seeps out, if over time, the spring would not start to run again. Yes, there are some big oak trees growing and dying near the spring, but I like to believe that 150 to 200 years ago that spring flowed freely; clear water at all times. The Native Americans must have known it was there, as did the buffalo. It would be a very interesting experiment for me. Now it only runs strong when it has rained. However, I will not cut all those trees because I like the vegetation, including the cedars growing up the steep incline. I know there is water and I don't even have to put my arm in the cracks.

August 3, 1994

Wheat Harvest '94

When I look at a loaf of bread, I see a whole year of farm work.

You work or live toward an event--the county fair or the annual wheat harvest. This year's wheat harvest is gone, long gone. The predictions are fulfilled or were proven wrong. Money was made or money was lost--maybe some money was made, most of it went to the bank. The harvesting stories are still told across a cup of coffee. How you just finished cutting the wheat and it started to rain or how a bearing gave out after the first round and you had to stop, having just started.

For me, this year's wheat harvest began one evening when the typical sweet, dusty scent of ripening wheat drifted across as I passed a yellowing wheat field. It seems a long time ago now, but that was the start.

The next few weeks you could see the combines raising dust, going around and around while trucks and trailers were impatiently waiting along the edge of the field. Every so often, the big combine would pull over and disgorge the golden, yellow kernels into the slowly filling truck. Once full, the truck would creep out of the field and be on its way to the elevator to unload and hurry back.

For a while, the freshly cut wheat fields would look neatly cut, like a G.I. hair cut with the clean stubble. However, it did not take long for the golden field to turn green and greener as the weeds took over. Some fields were quickly burned and they turned black until they were disked, or cultivated, and then they turned brown, the color of soil. Other fields were disked without burning and the wheat stubble turned a brown-gray-yellow as the stubble was worked into the ground. The fields losing more of the golden color over time, with the wheat harvest fading into memory.

In some fields, the loose straw was baled into large, round 1,000-pound bales which were moved to the edge of the field where they were neatly stacked in a row. I did see some square bales, but the small, square bales, although very handy, are not made often anymore. Too much handling involved.

If you drive through the state, that is about all you see as being left of the 1994 wheat harvest. Now farmers are preparing for the next crop and for some, another chance to hope. You see much emotion in the now freshly cultivated fields.

Yes, everything is under control at this moment, but each growing season is filled with anxiety. We all know the stories of crops lost a few days before the harvest could start. A friend of mine lost this year's crop due to a hailstorm. The second storm finished it off. And I know of other places where the crop was lost, never harvested, too wet, or the weeds took over.

Cropland is different than pasture. They both have stories to tell, but the story of cropland is often more dramatic. We make the decision of what to grow and then we prepare the land. The big machinery turns it. It lays fallow and soaks up the rain. When it gets to be time to sow, we need the right soil and weather conditions to make the crop germinate and then you watch it daily, and if not daily, surely weekly, and you see the crop ripen. In the meantime, you have worried about insects, weather, and you have watched the market, and you hope. Of course, farmers will always put crops in the ground, but I think it takes courage and faith.

What do you tell a friend or neighbor who lost his wheat crop when the same storm just bypassed you? When, after the storm, you drive out and see on one side of the road a tangled, flattened mess of wet straw, and on the other side, the heavy kernels slowly shaking the droplets of rain? And as you stand there, the sun peaks out and shines warm on your back, while in the distance the dark storm clouds move on, creating havoc in its narrow path. It is not easy. What do you say when you slowly drive home and your wife looks at you and without a word, she knows the crop is lost?

When you drive the state and see the empty wheat fields, remember they have stories to tell--of a golden harvest, happy faces, of loss and sadness, of faith and strength of hope. Whenever I look at a loaf of bread, I see a whole year of farm work. I see plowed fields. I see ripening wheat, and I have hope and faith.

August 16, 2006

The Best Crop of All

Kansas is known as the "Wheat State." A front plate on my wife's car, a cute RAV 4 Toyota, is attached in lieu of the dealer's tag. I wanted people to see that we are from Kansas. The black tag with golden wheat bending and the word Kansas does just that. Besides, when we tell people where we are from, they ask, "Kansas?" They look at us and wonder, with our accent. Hey, sorry about that. We are Kansans by a roundabout way. But we are Kansans. Look at our car--the tag. We are from the "Wheat State."

However, I must admit, ever since the last time we drove northwest and saw wheat fields in Idaho and Washington state, I have wondered about the claim. They grow a lot of wheat up there. And when I say, "up there," I mean it. On some of the fields I passed, I wondered if the wheat was sown by helicopter or if they just threw the wheat seed down the slope when the wind was right. There was some very, very steep country being farmed and quite successfully according to a friend of mine who lives in Moscow, Idaho.

Of course, besides wheat, Kansas is a beef producing state. We eat beef. We like beef--good beef. And there are other agricultural crops--soybeans, sunflowers, alfalfa hay, and more. We are an agricultural state with most of our land; I believe 98 percent in private ownership. When farms were smaller, farm families often lived close enough that one could see one another, especially in the eastern part of our state.

Today, as farms have grown bigger, especially the crop farms, the farmsteads may be further apart with abandoned houses and old barns and sheds standing in-between. The old buildings are slowly falling down, refusing to hurry. Every so often, you can see the old building pushed in a pile, where a farmer no longer wants to look at the old place or had no memories of grandparents once living there.

In my collection of Kansas photos, I have pictures of old farmhouses, some stone, with perfect straight rooflines but no window frames and no doors. The stones are squarely laid. Sometimes, the windowsills have some slight adornment, telling of better times. Often, over the door there is a stone with a date. We are now in the year 2006 and each year the history of the old stone house and falling buildings fades further and further into the past.

Whenever I slowly walked around an empty old house, I tried to envision the family living there. The parents, sons, and daughters all working together; tending the crops, the livestock, and each other--and don't forget the neighbors. They created and were a community. They helped each other and I expect still do. However, there are now fewer people as many have left the land. Bigger and bigger machinery has taken over where, at one time, many hands made light work out of hard work, monotonous work, and just plain work.

On my desk stands a black and white postcard. I bought a couple of them in the museum in Abilene. The card shows a team of white draft horses pulling a loaded hay wagon with loose hay. The horses are pulling the load uphill with a good, strong walk. They seem to enjoy the load and the opportunity to show off their strength. On top of the load, the farmer holds the reins in both hands. He knows his team and does not want them to run away with the full load and lose the hard work of mowing, raking, turning, loading by pitchfork.

On the back of the postcard, I read, "Harvest--this Pennsylvania farmer is carting home the hay at summer's end. The farmer's life was a difficult one, often requiring him to work from sunrise to sunset." The photo is from a collection in the Library of Congress. There is neither date nor the name of the photographer.

I like the photo for the power of the horses and the old way of hauling loose hay--the hard work, the many hands it took to get the work done. And bringing the load home was only half the work. Once home, it had to be unloaded and stored in the loft. But before that job, there always was the ride home and I remember riding on top of the load with the blowing hay sticking to your sweat and the cool breeze slowly cooling you. Sometimes we, as kids, would lie down and sink partly into the aromatic hay and watch the clouds drift over as we moved along.

Summer days were long days, but so were winter days when chores such as feeding had to be finished by lamplight. I don't think that a farmer's day today is that much shorter. True, the modern machinery makes it easier and an air-conditioned cab is very nice, as are cabin filters to keep the dust out, but the hours can still be long and tiring. And with modern machinery, you have to stay with it. You have to pay attention

My son-in-law, who grew up on a farm in western Kansas, made a remark which is interesting and very true. Our farm families have given us one more crop for which this country should be very grateful. Besides the wheat, the corn, the beans, and all the rest, they have given us their children for whom there was no place or need for on the farm or ranch.

Those children, the sons and daughters who left the farm, became engineers, veterinarians, doctors, mechanics, nurses, teachers, and whatever other profession they chose. They were hardworking individuals by nature, able to solve problems which they saw as opportunities. They were mechanically inclined. The farm and farming had trained them that way. They know responsibility and have a tremendous work ethic. It is true that some turned their backs on the farm and never wanted to return. But those parents, who made farming a way of life for their families, gave this country perhaps the best crop of all and still do, as their children leave the farm.

August 15, 2007

Toto

I have mentioned to you earlier that before we came to Kansas, we lived in Blacksburg, Virginia, where I was working for the Horticulture Department managing the Horticulture Research Station. It was an interesting job. We rented an old but unique farmhouse on the corner of Glade Road. The place was unique because as spring approached, all kinds of flowers came up out of the cold winter soil near the foundation of the house. I later learned that, at one time, a horticulture professor had lived in the house and that the woman had been an avid gardener.

Besides a tall blue spruce and other interesting shrubs, there was a small orchard with apple trees. The whole setting was typical of an old, small farm, including the old barn. We rented the old house with about five acres, which gave us room to keep a small Chincoteague pony we named Star for the star on its forehead and a flying seagull on its hips.

When we came to Kansas, we brought that pony with us. It died on our hill here at an old age. It had a good life and gave pleasure to our children. It was a Christmas gift to our oldest son a long time ago. What was interesting was that the pony was cheaper than a bicycle.

As Blacksburg grew, the old farmhouse and 35 acres were sold and a shopping mall was built. By that time, we moved to our own house further down Glade Road. That was another farmhouse, a beautiful house with a slate roof. We loved it.

But what I will always remember is that coming back to Blacksburg, Virginia, after we moved to Kansas, I found a Dodge City Steakhouse where our first farmhouse and barn once stood. It could have been a Pizza Hut or a Burger King. No, it was a Dodge City eating place. And here I was, fresh from Kansas. I decided to return that evening and eat a good steak where once our kitchen had been. I enjoyed a good meal in an atmosphere of branding irons and other typical cowboy paraphernalia like ropes, bridles, and halters hanging on the wall of the eating establishment.

On a recent visit, we stayed at a small motel along the Blue Ridge Parkway. It was constructed in the style of an old log cabin with wide and spacious porches on both sides of the seven rooms. In the evening and early morning, I would sit in a rocking chair and watch the sun rise over the hills, listening to the birds. It was very beautiful.

Walking into the office and waiting for someone to come and register me, I looked around and chuckled when I read a sign which said, "We are no longer in Kansas, Toto." Maybe it said, "I think we are no longer in Kansas." When you read that standing in a place hidden in North Carolina, you wonder. Later on, when their family dog came over to make acquaintance with me, I heard his name: "Come back, Toto." Now, I really was thinking that is funny. Here I am all the way from Kansas, pulling off the road in an out of the way place in North Carolina Appalachia and I am surrounded by Dorothy--funny.

I know that the story of *The Wizard of Oz* is part of Kansas and there is a yellow brick road to follow in several towns. But we seem to have exported the story to far-off places. When my wife was a small child, her father read the story of *The Wizard of Oz* to her and her young brothers. When her father read the children's book, he would dramatize the story by using his voice.

I have wondered where the old book, *The Wizard of Oz*, ended up. I would expect it was read until it fell apart. They read the Dutch translation, which makes me wonder in how many different languages was the book translated? Many I think, making people wonder if Kansas is where the tornadoes roar. And with the recent big tornado which wiped Greensburg off the map, the theme of the old book must have been rekindled as to what Kansas is all about. Luckily, that is not so, even though the tornado which swept Greensburg away was terrible!

But natural disasters are all around us, depending on where we are. For a long time, nothing happens, and then suddenly it strikes. The story of the children's book, *The Wizard of Oz,* tells us that it can happen. Greensburg shows us that it does happen. And reading a sign in a faraway place about Kansas reminds us where Kansas is and that is a good thing. I hope other people will read the sign and decide to drive out west and come look for themselves to see what Kansas is all about and as we used to say, "linger longer."

February 3, 2000

Kansas

The February-March issue of *Country* magazine has an article by Christy Kass who now lives in Pennsylvania. No matter how hard she tried to explain it, city folks could not appreciate her love for Kansas.

Last week, I gave two talks. Both were accompanied by a slide set of photos I have taken over the many years that I have traveled throughout our state. The first group was World Friendship. It is a group of people, mostly women, from many different parts of the world whose husbands are pursuing an advanced degree at Kansas State or are members of the faculty. It was my wife who twisted my arm and said, "You can do it and it is Kansas Day this week and a program on Kansas would be very timely." I let her talk me into it and it was a very rewarding experience. Many of the women do not speak English or only haltingly, but they enjoy being together.

It is true that many of them, for the duration of their husbands' time of study, don't get beyond Manhattan. So, when they return to their own country, their image and knowledge of our state may be rather limited. That is also the reason why I let my wife talk me into accepting the talk. The old Extension motto, "have slides, will talk," took the better of me.

I had to chuckle when my wife reminded me to speak clearly and not too fast, or else they would not understand me. The old saying "In the land of blind people, the man with one eye is king" applied here. I spoke slowly and clearly and showed them Kansas, rural Kansas. I went from the east to the west and from north to south. I showed cropland under cultivation, hayfields being baled, and cattle grazing. I shared the uniqueness of many of our old barns, one of which showed a definite eastern European influence with its bell-shaped facade.

Among the women were those who came from Asia, South America, Eastern Europe, and Russia. As I showed slide after slide, watching their faces as I clicked through, the murmurs of "aaahhh" could be heard. The snow scenes showed cold weather and empty fields. There were scenes of plowed fields with only a touch of snow, creating beautiful patterns drawn by farm machinery. The fence lines with down-turned boots stuck on the posts made them laugh. They saw the humor. The women also laughed at the sunflower painted around the circular trap door at the base of a large elevator complex. When I showed golden cloud formations with the sun shining through from behind the clouds and silent sunsets, they sighed even more.

As I shared our state with them, I wondered if some scenes reminded them of home, a home to which they most likely will be returning following the completion of their husbands' studies. I had no time to ask them because the program was only 45 minutes.

Later in the same week, I was asked to give a 55-minute talk to young farmers at the Farm Bureau Convention held in Manhattan. The topic I was assigned was "Farmstead Landscaping and Culture in Agriculture." The latter part I had tagged on because I knew that in the 55 minutes allotted, I never could do justice to farmstead landscaping and not only that, years ago I wrote a publication on Farmstead Landscape Design and anyone seriously interested could ask for it as well as a set of videotapes on the same subject.

Because of that, I chose to elaborate on culture in agriculture and use the same slide set I used earlier in the week with the agricultural and Kansas landscape scenes. I had the satisfaction that as I presented my program, more and more people drifted in. The group was made up of mainly young farmers and wives. I showed scenes they were only too familiar with, but they never had looked at from an aesthetic viewpoint of being beautiful scenery or structure. I did hear a few soft comments of "wow" or "aaahhhs."

I knew that because of all of them being farmers, people who work the land, there was a risk giving this kind of program, but I wanted to touch them. Actually, I wanted their land to touch them. I wanted that for them and for their young families.

I finished this program reading a short legend. It is a legend from a West African tribe. The legend is called "The Sky Maiden." It so happened that people of the tribe noticed that their cows gave less milk and it looked as if someone had milked them. So, a young man of the tribe volunteered to stay up to see what was going on. After several hours hiding behind the bushes, he saw a beautiful woman descend from the night sky. She rode a moonbeam down from the heavens holding a milking pail. Once down, she gingerly stepped among the cows and started to milk them. Having filled her pail, she rose and climbed back up to the moonbeam in the sky.

The man was astonished and could not believe his eyes. But he knew what he had seen was true and it explained why the cows gave less milk. He decided to capture the woman. The next night, he set a trap and sure enough down came the woman again. He sprang the trap and caught her. "Who are you?" he asked. She explained that she was a Sky Maiden, a member of a tribe who lived in the sky which had no food of their own. And it was her job to come down to Earth each night and gather food.

She pleaded with him to let her go and as always is in these legends, she would do anything he asked. The young man said, "I will release you if you promise to marry me."

She said, "I will marry you, but first you must let me go home for three days to get ready. Then I will come back and be your wife." He agreed and set her free. Three days later she returned, this time carrying a large box. "I will be your wife and will make you very happy," she said, "but you must promise never to look in this box."

They lived happily, until one day, as his wife was away, the young man was overcome with curiosity and opened the box. Opening the lid, he saw there was nothing in it. When his wife came back, she realized that he had looked in the box because he still looked puzzled having found the box empty. "You looked into the box, didn't you? Now I can't live with you anymore," she said.

The man asked, "Why? What's so terrible about my peeking into an empty box?"

"I am not leaving you because you opened the box. I expected you would sooner or later. But I am leaving you because you said it was empty. That box was not empty. It was full of sky. It contained the light, the smells of my home in the sky, the scent of rain. When I went home for the last time, I filled the box with everything that was most precious to me to remind me of where I came from. How can I be your wife if what is most precious to me is only emptiness to you?" she replied.

After my talk, a few of them stepped forward and shook my hand saying, "That was different." They had seen their land, even recognized some places, but they had seen it with different eyes.

The Farm-Our Land

Music in the fields.

December 1, 1993

Seeing an Andrew Wyeth Painting on the Farm

Oh, it cranked over and after several tries, it would just poeke, poeke, poeke....

I have always liked the paintings of Andrew Wyeth. I like his attention to detail, but above all, I like the feeling I read in his paintings, especially those of rural, winter scenes and old farm buildings. Now I am not an art critic, I simply go by gut feeling and how long I can look at a painting before moving on. I had to think of his paintings while cutting wood this past Saturday.

Mind you, I had no intention to cut any wood this Thanksgiving weekend. The plans were to build a half-mile fence up and down and along the hills while the family was home for the family dinner. We had it all planned, my sons and me.

Thursday, my youngest son and I set the last of the steel posts. I showed him how to line them up, sighting with the eye. Thursday was still a pleasant day to work outside, especially in the lee of the trees.

Friday was planned for digging post holes with the tractor. That day the weather was colder, but still good to work, at least it was dry. There was some snow on the ground, but nothing to worry about except perhaps the tractor coming down the steeper hills. But we never got to try because the old John Deere refused to start. Oh, it cranked over and after several tries, it would just poeke, poeke, poeke, poeke, poeke, poeke along. But it never really fired up. We tried everything but taking the carburetor off. We came to the conclusion that the carburetor was frozen. And as cold as it was, with the wind blowing across the middle of the alfalfa field where I had parked the tractor to keep the darn packrats from it, I knew that the carburetor would remain frozen. The rest of the day we hauled the heavy Osage orange fence posts in the back of the 4-wheel drive pick-up my oldest son had borrowed and dropped them off where they were going to be part of the fence--corner posts, stays, gate posts.

I was glad we had a 4-wheel drive truck to climb up and down the steep hills. My only fear was to set the rangeland on fire with the hot muffler. We did not. But I could just imagine what a roaring fire it would be--dry, tall grass and the cold wind. The little bit of snow would not have slowed down a grass fire. At least we accomplished something that day, but not half as much as planned.

Saturday, I had things organized. We threw in a plastic pipe to extend the muffler and blow hot air on the carburetor and melt the ice. Hey, it worked. We got the carburetor really hot and water drained from it. But do you think the tractor would start? Forget it! Poeke, poeke, poeke, poeke, poeke--was all that it did.

When I was younger, I used to get very, very angry when these things happened. I am getting better at controlling my temper, especially in front of my sons. Not that I am perfect--no way--I would not want to be. But, when I realized that the darn tractor still would not start and snow was flying around, we all turned our backs to the cold wind and I said, "Let's cut firewood."

So, we drove to some good oak and ash which had fallen over and started up the chainsaws. It was while cutting up and splitting the wood that Andrew Wyeth's paintings came to mind. We were working along the woods--big snowflakes blew strong among the trees and partly covered with fresh snow, stood an old wheat harvester, a binder. It is like one I used to ride as a youngster, steering the horses while someone else sat to the right laying off the sheaves of cut and tied wheat. The old, rusted piece of equipment must have stood there for years. It was that old binder, the dark woods, the big trees, the drifting snowflakes which made the picture--very peaceful, very rural, and very real.

The three of us loaded two pickup trucks with good, hard, split firewood and called it a day. Sunday afternoon the sun was out and it was pleasant outside. My son-in-law and I drove out to see if he had any ideas why the old John Deere would do no more than poeke, poeke, poeke. We looked at everything once more, but this time looked also at the air filter. When I took the top off, there was a mud dauber's nest partly blocking the airflow. When I took the oil bath off, it was solid ice. Water had settled in with the oil and had frozen solid.

When we had it all cleaned out, Jeff said, "Now try it. I'll bet it'll go, old man."

And it did. The old tractor roared to life, blue smoke belching from the smokestack. After a while, the smoke cleared and the tractor purred like a kitten, no longer going poeke, poeke, poeke, poeke. It ran beautifully and I let out a big holler, "Yo, ho!" But you know, I still remember that scene on the farm that looked like a painting of Andrew Wyeth's. I never would have seen it if the tractor had started!

June 30, 1999

25 Year Storm

Without a calendar to remind me of birthdays, I am at a total loss. It is my wife who gently reminds me when there is a date I should not forget. Of course, my immediate family is not too bad. At times there is not enough time to send a card, but then they are only a phone call away.

I called my oldest son last evening to wish him a happy birthday and to let him know that 41 years ago I had been doing, more or less, exactly the same as I did on his 41st birthday--repairing flood damage. The rain gauge on the farm showed a good two inches. In the paper, I read that some parts of Pottawatomie County received up to seven inches of rain. They called it a 25-year storm. I'll be 92 when the next one comes.

My two inches was plenty, and it must have come down as a soaker because the beaver dam washed out, taking out some railroad ties I used to cross the spring, down the creek. Some large trees, which I planned to cut up and haul out of the big creek, were swept away. So, I don't have to do that job anymore. The creek crossings eroded and now need more work to get from one side to the other and to top it all off, the fence washed away at two places where water rushed through under the wires, but logs jammed it all and caused the fence to be ripped out.

Of course, my story is only interesting because 41 years ago when my wife said, "I need to go to the hospital, the baby is coming," I was doing similar work: tending cattle and mending fences. Only it was on the other side of the world in Australia where, as a young man fresh out of college, I managed a farm.

It shows you that farming all over the world is a risky business and that the weather is always the big unknown and it can play havoc with the best-made plans. It also shows you that maintenance--just the upkeep of fences--takes time and time costs money. My job of fence repair would have been easier if someone had been there to help. Many farm jobs go quicker with an extra pair of hands. But I managed very nicely and actually enjoyed the task.

I know it will all wash out again as long as my neighbor, who has the watershed, has dead trees lying around. The flood sent them crashing into our shared fence. While working to untangle the twisted, barbed wires, I remembered where the work gloves my son gave me were lying--back home on the workbench. Dipping my hands in the running water, I washed the blood off caused by the vicious barbs. I expect another tetanus shot won't hurt--every farmer should stay up to date on that one!

I remember 41 years ago, we had a cloud burst upriver from our farm in Australia and when I returned from the hospital, I jumped on the tractor and raced down to the river and was able to pull the big electric pump up to dry land only in the nick of time. That was 41 years ago, and I can still see it all. That storm caused a lot of damage up and down that river--the Hawkesbury River.

At that time, I had a cattle dog, a blue heeler who always kept me company when working on the farm. Now, many decades later, I have Kansa with me, our German shepherd. He kept a close watch on me as I splashed through the water while splicing wire and driving posts. In lieu of a hammer to pound the fence posts into the ground, I used a big rock, which worked fine. My only fear was it would break in half as I was hammering away. It never did--good Kansas rock!

When the job was finished, I dragged the tree stumps high enough upland, out of the flood path; leaving the pounding rock where I could find it for future use. On my walk back to the truck, I crossed the prairie hay meadow, now in full bloom with prairie flowers--brown-eyed Susans, fleabane, and the soft magenta of prairie alfalfa and bee balm.

Back home, my wife asked me a funny question, "Was it frustrating having to fix the fence?" She probably asked me because it had happened before and as I said, it'll happen again. But the task was simple. It had to be fixed.

While working, I listened to the gurgling stream. At least we have enough grass and water for the cattle and the prairie is beautiful. I enjoy it still just as much as 41 years ago--to take the good and the bad. Who needs a vacation when you are having fun!

May 22, 1991

All is Well on the Range

The last few weeks have been busy. All over the Flint Hills, cattle have been taken to pasture. On the country roads, you see cattle trucks and stock trailers loaded with cows or empty trailers hurrying back for another load. Across the hills, you could see men on horseback herding cattle or checking fences. Pastures have now turned green and cattle don't have to graze too long before they are satisfied, lie down, and in a dreamy way, chew their cud.

The first days, when cattle are turned loose on pasture, it seems they have to find the boundary for their range. I have watched them hurrying along the fence line, kicking up their heels or butting each other in a playful way. Then racing down the hills with papa bull slowly catching up; his stoic facial expression seems to say, "Settle down cows and heifers." At times, cows in different pastures will meet each other across the fence. For a while, pleasantries are exchanged while bulls give a low moaning bull groan, sometimes roaring into a more bellicose sound.

I enjoy seeing the different breeds of cattle. The Black Angus look so well dressed in their black business suits. The Charolais is all white. Although they look grass-stained from the new grass and if they lay down in the mud, they can look dirty. The Swiss Brown Gelbvieh looks a bit out of place. I think of the Swiss mountains with these cows and miss hearing the big cowbell of the lead cow.

I am partial to the Herefords, liking their cheerful white faces and the deep red of their coats. Being fully aware that my appreciation of cattle seems to be based on looks, only a rancher might say, "Hey man, there is more to cattle than looks. Some handle easier, some are better mothers, some have fewer problems calving and others are quiet grazers, gaining weight easily." Ranchers, especially if they run steers, may have a mixed batch with some dairy breed steers thrown in.

The reason I am partial to the Hereford may be that as a boy I once cut out a large Marlboro advertisement. It showed a suntanned, tall cowboy standing on a hill looking over his cattle. The reins of his horse hung loosely over his arms while he lit a cigarette, as his keen eyes glanced from under his broad-brimmed hat over his cattle. All Herefords were happily grazing below the hill. I cut that picture out very carefully and glued it to the wallpaper right above my bed. Those cows were the first thing I saw when I woke up in the morning. It looked strange to me--the wide-open country, riding the range. I was used to small Dutch pastures where only well-behaved dairy cows grazed.

The power of advertising--the suggestion to smoke like the Marlboro man--could not influence me. I was still way too young to smoke and simply not interested in it. But I remember looking at that scene of white-faced red cattle and the strong sturdy horse. Young boys can and should dream. Now walking or riding on our own small ranch, I sometimes have a flashback to that scene I glued to the wall above my bed many years ago when I was a mere boy.

While we are still renting the land out for grazing, we don't run our own cattle and what we run is a mixture of good-natured cows and calves. So far, we have had wimpy bulls and one mean bull running the cows. The mean bull, an Angus, was cocky and smart. He found a hole in the fence and once out, he liked to play games of "hide and seek" when I would try to drive him back. While I was walking, pushing him along, he would trot ahead and hide behind a bush and wait for me. I often wished I had a horse. Once I fixed the hole in the fence, I didn't have any more problems. My son did not like that black bull either. The bull chased him once while he was riding his pony around the herd. Anyway, that is what he said.

Things are quieting down now. The cattle are settling into a routine of grazing, drinking, walking up to the salt lick and the calves are gaining weight. The bull does what he likes to do most. The grass is growing. All is well on the range.

January 10, 2007

The Old Seat

Once more, it was snowing in Denver. That would be the third storm if indeed the accumulation would make it a storm. Much of that would probably depend on if the wind would come up and send the snow flying. For Manhattan, the weather forecast was possibly some flurries for the weekend.

Driving out to the farm I could see a change in the weather, dark clouds in the northwest, but nothing threatening. Actually, it was a good day to walk the fence line and check where work needed to be done. Over the years, the fences have been improved upon dramatically; even though a cow or two still wandered off to what they thought were greener pastures. Actually, all it may be is just a cute bull on the other side of the fence. However, if that is the case, it often is the bull that will come across to claim his love and do the honors.

But during the winter there is no stock on the hills and when walking the range, I make a mental note where the sumac is too invasive and we must spray, and how best to burn when the burning season is here later in the spring. A good burn is always a challenge, but it surely can improve the stand of grass.

However, this morning I had the fence line in mind and in particular, the short fence near the gate we use to put the cattle on and take them off the grass. One of the gate posts had broken off. Someone may have pushed it when entering as the old gate is kind of narrow--too narrow. Mind you, it is just a wire gate so it should be no problem to widen it. All it takes is to place the posts--stout posts--further apart, making sure that trucks can easily back up to it as well as drive through.

I do not like trucks driving on the range but at times you need to. Just like a few years ago, my neighbor's bull came on our land and it happened that at that time my renter's bull was attending to the cows. The two got into a hell of a fight, which my neighbor's bull lost. My neighbor had to drive out onto the range to pick up his bull. I felt bad but these things happen. That is why you want good fences!

I noticed some stout Osage orange trees growing along a woodland edge. They'll make good straight gate posts. With a good stay, that gate will last for years--way beyond my time! Now that is a thought!

Pulling out the one crooked but sound corner post, an Osage orange post, I then also cleaned the corner where buck brush and rough-leaved dogwood have made a bird haven. With the leaves dropped and blown away, I noticed an old "No Hunting" sign, as if that would have stopped anybody from trespassing. On another post, a "No Trespassing" sign still hangs. Over time, both signs have been used for target practice. There must have been some random shots and a bullseye where the bullet pierced the center of the "o" of the word "No."

Leaving the "No Hunting" sign resting in the weeds, my eye caught the old, iron seat half-buried by dried leaves. I pushed the brush aside and pulled the old seat out. Holding it in my hands, I started to wonder. Who threw it in the corner? Who sat on it last and when was last? When did it break off? Because turning it over, I noticed that it had snapped off. There was a clear break. I turned it over again in my hands and studied the seat. Metal, with two rows of thirteen holes, each three-quarter inch to one inch round; thirteen, the unlucky number.

Whoever sat on it last had a very uncomfortable ride as the seat no longer held its original shape. It was somehow slightly inverted as if a very, very heavy man had sat on it. In the center where the bolt still held onto the broken piece, it had pushed up. It seemed to fit our anatomy, but not too comfortable if you imagine sitting.

My next question is from what piece of equipment did it break off: a cultivator, a corn picker, or a binder? I have several of the old machines leaning and growing into trees--big trees by now. However, all those old farm machines are resting along the edges of fields, bottomland where at one time they were used. This seat was lying near a corner of the range. Closest to it was a hayfield, prairie hay, and that may be what it was a seat from--an old mower pulled by two draft horses. That is the best I can come up with. Now think about the moment that the seat snapped off--oops!

As it happened, one would pull on the reins when tipping sideward or backward. The sweaty horses, glad of the sudden pull on the reins indicating stop. Maybe the pull was more severe than usual, putting pressure on the bars. But they stopped, the driver looking down in disgust at the broken seat.

Through one of the holes in the iron seat, a piece of wire had been twisted as if to hold or hang something. Could it be that as he mowed, a can of water had swung under the seat or maybe, as the seat was so uncomfortable, he had tied an old, folded sack or cushion on it to save his hemorrhoids? Because without some kind of protection, I am sure a hemorrhoid or two would be nurtured. Thinking about that, I remember how we pushed hay into a sack and placed it on the iron seat to soften the bounce and punishment. It worked.

Right now, the old seat lies in my pickup truck. I think I will take it to my son who wanted one to make a seat for his workshop. He can weld a stand and fix it. But remember--watch your hemorrhoids!

January 15, 1992

Rich Farmer, Poor Farmer

My trouble was I played too intensely. To think this was just a family game.

New Year's Eve seems a long time ago, even though we are only fifteen days into the New Year. There are still reminders of the holiday season. Here and there you still see Christmas decorations displayed. Some city parks still have the leftovers of Christmas trees piled together and single trees can be seen lying in the alley or forlorn on the lawn, sometimes a forgotten silver ball or tinsel sway in the wind. There have been drives to collect Christmas trees to be used to improve fish habitat in the lakes. The trees tied together and weighted down are to be dumped at selected sites of our lakes.

Our tree is now standing outside and the birds enjoy it. I'll let it stand for some time and will take it from the holder and let nature take care of it in our woods. Over time, the needles will fall off and branches will turn brittle and break off. Walking the woods, I'll watch the decomposing process as I pass the old tree, seeing it get smaller and smaller until only the main trunk and stump are left.

Glancing at it, I'll recall the brightly wrapped presents which lay under it. But more importantly, I'll recall the burning fireplace, the hot coffee, and the time we played the family game "Rich Farmer, Poor Farmer." It is quite an educational game. If you have never heard of this game, I can best explain it as similar to the Monopoly game, which most of us have played one time or another. Only, Monopoly is structured around houses, streets, and real estate development, which you can lose or turn into a fortune. Of course, paper play money is used for all the transactions. The same with the game "Rich Farmer, Poor Farmer." However, instead of buying real estate, you buy farmland, cornland, peanut land, cotton land and you improve the land. You plow it and raise the value of it. You fertilize it and raise the value more and so on through planting and harvesting.

Four of us played the game. One of us was the banker and handed out an equal amount of money to each player. That was our working capital and then we turned the spinner and moved our tokens, little tractors, accordingly. Whenever we stopped on "land not yet sold," it was quickly bought and in your mind, the cornfields were brought into production through plowing, fertilizing, planting, and finally, harvesting. Each step meant more money invested, but the return came when another player landed on your piece of land. He or she had to pay much more for the improved condition.

The variables were there: the cards of opportunity; your tractor or combine broke down, pay five thousand to have it repaired; you suffered a hailstorm or flooding; don't collect payment for a round. Sometimes you made some money when you pulled a card with "You sold farm machinery at an auction, you get five hundred dollars."

We played for several hours and did not even finish the game that day and continued the following evening. Once all the land was sold, the game revolved around the four of us improving our land and paying our debts. As long as the money came pouring in, there were no problems. But once money got scarce, players had to remove improvement or even sell off their land. To the four of us who all love land more than real estate, the game had moments of pain. At least I felt it when I had to sell my land to pay my debt to the other players.

Quite early in the game, I noticed two things. One was that the bank seemed to get our money one way or another and the other was luck. Pure luck had much to do with who won the game or who got the breaks, such as good weather or not too many mishaps. With luck, you were able to build up capital and if good fortune stayed with you and you did not land on other people's expensive land, then you made it. If they owed you and had to sell, you could buy up their property and you would get richer and richer.

My trouble was I played too intensely. I improved my land. I tried to be progressive yet hold onto some working capital so I could pay my debts. But I ran out of money and had to start selling land. It still hurts me that I lost my first peanut farm and then my corn farm. To think this was only a game, "Rich Farmer, Poor Farmer."

February 3, 2010

The Oak Tree

The huge oak tree stood for a couple of years like a windblown, wind-torn umbrella. You have seen those umbrellas when in rain and thunderstorms the opened umbrella goes--whoop--and in a second is turned inside out. One moment there was some protection from the sleeting rain and the next moment, rain beats in your face. You might as well toss the broken umbrella in the trash as it has become useless.

It was the same with the mighty oak. Ever since we owned the land, I have enjoyed that oak. It stands on the edge of a hay meadow along the woods where a draw comes down the hills. Upstream from the small trickle, a dam was built which holds a lot of water. Seepage provides water to the tree. I have measured the tree. At chest height, it is fourteen feet, actually 174 inches, in circumference. I don't know if it is the biggest oak on the farm, but it surely is one of the biggest.

For weeks now, a group of young guys has been cutting up its limbs into blocks which then are split, as the tree has cracked wide-open from the base to the crown. The first loads were easy. It was cutting the ends, the tips of the branches, rrr--rrrr--rrr. But slowly, the branches got thicker and thicker and cutting became trickier. But they managed.

It made a difference that they were friends and could laugh together. At one time when I walked up, I heard several chainsaws roar and smelled smoke from a small fire in the snow and the aroma of cooked hotdogs. They had been hauling the split wood out and had it stacked in neat piles. I don't know if they all cut together or each cut for himself. I didn't ask. Having given them the tree so the trail would be cleared, they gladly took it. I enjoyed their laughter and camaraderie.

Saturday, my grandson went along to help his brother-in-law. They had a wood splitter to make splitting easy, or should I say, easier. Because by now the major branches were so thick and knotted that even Paul Bunyan would shake his head, sit down and look at the job at hand to see if there was not an easier way of making small pieces of split firewood out of those large, solid blocks the size of tables.

The reach of the branches was so far that they had brought wheelbarrows to fill and push to the trailers and pickups. The scent of fresh-cut wood hung around and was on their clothes.

As they worked over several weekends, I had told them what my rules were as far as driving across the land. To get to the tree they had to drive across the creek, dip down a gravel base, and up the bank. Then they had to drive across three hay meadows. As long as the ground is dry or frozen solid, there isn't a problem. Even with snow on the ground, it is okay, being the only way to get to the tree.

However, generally, I do not drive over land--hay land or whatever. So last Saturday was eighteen degrees Fahrenheit, no problem. But as the day warmed up and the sun came out, it did start to thaw out on the surface. Deeper down, the ground stayed frozen. It worried me because a few weeks ago they went in with full might--trucks and trailers--but when they came out, the ground was soft on top and they made tracks. Their 4-wheel drive trucks have big, ground grabbing tires--macho. Their deep ruts stood out--dark brown-black in the last of the snow. They made deep tracks. In the bottom of the creek stood a loaded trailer they had not been able to pull up the creek bank. Oh, what a mess!

They knew I would be upset, if not mad. I don't need that. The land does not need that. But I knew it would be better to get the stranded trailer out sooner rather than later. In a good rain, the creek runs several feet deep. In a torrential rain, it will run eight feet or more. And I have found cans floating over the banks and dumped on the middle of the hay meadow.

I told them to get the trailer out early in the morning when the land was frozen. When I looked a few days later, the trailer was gone. The creek once again rippled across the shiny flint rock as it has done for thousands of years. Only there were more tire ruts going up the bank--and deeper. I know, sometimes it just can't be helped. But I don't like it. It hurts the land. Ruts cause erosion and erosion destroys.

True, the ice storm destroyed the huge Bur oak and with it a whole ecosystem that went on in, under, and around the tree. The red-tailed hawk which used to sit in the top no longer has a perch from which to snatch a mouse that feeds on roots and acorns or a squirrel that feeds on acorns. No longer is there the mole or the snake which feeds on mice and snails, nor birds on insects.

The web of life, a food chain from plant life to a red-tailed hawk that only preys on small animals, is gone with the loss of this tree. The Bur oak was only one of many trees and life will go on. In the spring, acorns will sprout. I will protect some descendants of the oak so that they may grow and someday they will be mighty oaks. With the huge oak gone, there is now light, and light helps things grow.

I'm looking at the stump of this fallen tree. If I can cut it off, say ten feet above the ground, maybe I'll build a treehouse on top. That would be a nice retreat, a treehouse with a hammock. I like that. I can see it. You can hear the rain fall on the tin roof and see the snow blow by. Now I need a small stove. I could become a hermit living on the mighty stump of what once was a mighty oak!

April 20, 1994

Prairie Burn 1994

We drove home in the dark, watching other hills still slowly burning with necklaces of fire around the dark, distant hills. This past week, the air has been filled with smoke. There hangs a haze over Manhattan, for that matter, over all of eastern Kansas. At night, with the windows wide open, the acrid smell of burning grassland enters the house, the bedroom.

When I drove out to the farm this past Saturday afternoon, all I carried with me was the old rake, a long-handled shovel, a few buckets of water, and a box of matches. I was not planning to burn, thinking it still was too early. However, I must admit I was getting anxious to burn and had been watching the stage of redbuds blooming as an indicator for the optimum time to burn.

But the redbuds are taking their time to bloom this year. Only now are they getting to full bloom--a good time to burn prairie. But to burn, the weather has to be favorable, especially the wind, which is a big factor in a successful or disastrous burn. The last thing you want is a fickle wind shifting on you all the time once you have set the fire.

On the way out to the farm, I noticed a lot of smoke from that direction, but upon coming closer could see that the burn was behind the farm. No one had told us that they were going to burn that Saturday.

At the farm, I met up with my oldest son and his helper, my grandson, Kit. Together we drove to the back of the farm and walked up the hill. From there, the smoke was much closer and before long we could see a long row of fire coming our way. I looked around for anyone who was attending the fire but saw no one.

Slowly at first, then with a burst of flames depending on fuel and the wind, the flames continued rolling forward, coming closer, crossing the fence. Looking around again, I still did not see anyone.

All I could guess was the fire started northeast from us and now was moving southwest, coming straight to our property.

The decision to burn or not to burn was made for me regardless of the bud stage of the redbuds. The three of us quickly walked back to the pickup truck and drove around. We drove up to the west edge of our grassland and parked the truck in the green alfalfa field. I got out a few matches, lit them, and with the rake pulled the fire through the grass along the wooded edge, a natural barrier. The immediate heat of the hungry flames made us keep our distance. But as the flames roared up the hill to meet the advancing burn from the east, the big flames leaped away from us. There was no stopping any more now. We were committed to the burn.

I quickly pulled the starting fire around and then walked back to where we began near the big pond and started a fire line going northeast. By now, the invading fire had met our backfire and where the two met was a perfect triangle of un-burnt grass on the slope of a hill. We watched the triangle get smaller and smaller as the fire raced toward the center, fueled by last year's un-grazed range. That clean burn is the dream of a good prairie burn. If you can circle the fire and let it burn out at the center--perfect. But it seldom happens that way.

I threw a few more matches in the dry fuel and continued to pull the fire east. Again, the flames quickly got away from us, roaring and crackling up the hills. By now, we had given up beating out the backfire, letting the woods do the job. However, I never like to do that.

I see a prairie burn as a clean fire, but when seeing woods on fire, I squirm. The dead branches and stumps smolder for days. A hollow tree goes up like a chimney. I just don't like it. I hate to burn the litter of old, rotting leaves which make good humus and help absorb the rains when they come. With the recent rains, the woods burned slowly and later on we quickly put those fires out. By that time, the entire prairie had burned clean--our "spring rite" was over.

In the early evening, my wife drove out. She climbed her favorite hill which had not burned. From there, she could view all the blackened hills on the other side of the creek. My son, grandson, and I sat wearily on the tailgate of the pickup truck, reflecting on another successful burn. I have heard enough about controlled burns and talked with enough ranchers who admit that they dread the burn--but it is a necessary management tool, so burn we must.

My son drove home. My grandson held the bleached skull of a cow on his lap. My wife and I attended to the last spot fires along the woods. By now, the air was cooler, the humidity picked up and fires doused out. Soon we too drove home in the dark, watching other hills still slowly burning with necklaces of fire hanging around the dark, distant hills.

June 3, 1998

The Old Lady

Walking purposefully, she crossed the dirt road in front of my old, blue truck. She looked like a middle-aged lady having gone shopping for shoes, found what she wanted, and now was on her way home to prepare dinner. That is what she looked like, all prim and conservatively dressed. When I saw her, I did not know we would meet her again later that evening.

I had gone to the farm to see if Dean and Bud had time to cut the brome. It was getting ready to be cut--maybe by leaving it stand for a little while longer we might get more bulk, but the protein content should be high right now so I would say cut it. Opening the farm gate, I saw that indeed they had cut the brome. It already had dried some, so I expect they had cut it earlier that morning or the day before.

When you enter the farm, the farm road leads you along the wheat field on the one side of the road and a prairie hay meadow on the other. Halfway, there is a small knoll where we always park cars, trucks, and tractors. Years ago, Bud threw a few handfuls of brome seed over the knoll and over the years, the small stand has established itself. And yes, small as it is, we mow it and once cured, we pick it up with the baler--every little bit helps.

I parked the truck, called the dog, and walked down the farm road to the brome fields. The dog chased a squirrel up the tree where it sat, snickering at him. Kansa put his paw up the tree trunk and his big bark reverberated through the trees. I am glad he can't climb. If he could, there would not be a squirrel left on the place.

As I moved on, he came running after me, then passed me to take a swim in the beaver pond. There are no beavers at the moment, but there was at least one active this winter. However, I think he moved on. There is still water behind the dam and Kansa likes to go splashing--boy, he stinks when he comes out of it. I wish the dog could shake the stink out of himself as he shakes the water.

Generally, he gets rinsed down once we get home.

All the brome fields had been cut and the swath of green hay lay colorful on the contour of the fields. I flipped some hay over to see how dry it was underneath--it was getting there, one more day. I hoped it would not rain that evening and if it did, would just be a sprinkle. There was talk of rain and yes, we could use it on the soybeans, alfalfa, and hay grazer. Farming is a crazy business. Like Bud said this evening when I told him the story of meeting up with the old lady I talked about earlier, "You can't win them all." It was three days ago that I saw the old lady.

I never finished that story but it does have an ending. That evening, after having walked the hay fields, I walked back to my truck. By that time, the sun was setting behind the hills and the dog was just walking beside me. He had done all the running and chasing for that day. When we got to the truck, he got in and lay down.

Looking to the west to the setting sun, I noticed that a very small patch of brome hay had not been cut. It was an odd-shaped triangle, no more than ten by twenty feet. It made no sense not to cut it and was not something Dean would do. Curiosity got the better of me, so I walked to the patch to see if there was a reason why it was not cut--the small patch was leaning against the wheat.

I walked toward it into the setting sun and got closer. Just looking down, I saw this big, dark form crouching in the tall grass. I could make out neither form nor shape, but I saw it move--it was alive. I stepped forward, closer. I was just putting my foot down when, all of a sudden, this dark form exploded in my face--a big hen turkey flew off her nest. It all happened so quickly that it did not scare me. It was the old lady who I saw crossing the road earlier, but who would have thought that she would build a nest so close to the farm road? I counted eleven eggs.

The little old lady had flown away. Hoping she would come back I did not touch anything, but I now knew why Dean had left that odd-shaped triangle of grass uncut. I expect the old lady sat on her nest until the last moment and then jumped in his face too. Dean said later that when he roused her, she flew up and away and as she flew, she dropped an egg--so she laid one in flight. I hope that if she does not have young hatch out of this clutch of eggs, she has time to find another place on the farm to start all over. At least we have turkeys on the farm, but we can always use more--real turkeys that is!

July 8, 2009

The Weather Forecast

A few weeks ago, I talked about checking on the horses with a friend and we got caught in a bad thunderstorm with fierce lightning and streaming rain on top of the hills on the farm. Not very smart, but we had to check one of the horses which was badly lame the day before. I was afraid of an abscess in her hoof. All this was before there was another storm and my horse Flash got killed by lightning.

Anyway, as we haltered the mare and checked her over, the center of the thunderstorm caught up with us and my friend lay down flat in the grass. A wise move in regard to the lightning flashes and strikes close by. I decided to lead the horses down the hill into a small ravine. There I examined the lame leg and foot further until the horse spooked and we decided to let her go as no further treatment seemed necessary.

We then crawled under a dirt overhang which had eroded over time. Now get this. While we sat there, squeezed under the narrow ledge with lightning striking and flashing, with big, rolling booms and rain pelting down, washing mud off the banks, my friend took out his cell phone to call his wife, who would be sitting behind her computer, to ask her how long the storm was expected to last. Remember, we are a fair way from our parked truck and had to walk across the top of the Flint Hills to get where we needed to be. Here we sat. Rain was coming down in sheets and changing directions so that our safe place, however safe it was, started leaking and the narrow rivulet started collecting water and began to run.

His wife did not answer, so he called another friend who picked up. "Could you tell me how long this storm is supposed to last?" my friend asked.

The rain was pelting down and lightning crashing. Upon hearing the noise, she asked, "Where are you?"

Answer: "Oh, we are sitting under a ledge in the middle of a pasture off Bennett Road, just west of Wickstrum's place. It is raining and thundering hard and we are off the flat top."

Apparently, she knew what button to push on the computer as the answer came through. "The storm should pass you and be over in about twenty minutes."

"Thanks," my friend said.

Another twenty minutes. From where I sat and knowing the lay of the land, I saw the storm make a circle and come back with a vengeance. With more water seeping down the bank, I felt that we were losing it and I saw no let-up in twenty minutes. We decided to make a break for it. We rose and started climbing up the hill. Kaboom--boom--boom, the thunder went.

I traced our path as much below the ridge as I could. However, we had to cross it. The rain and wind pushed us in the right direction. Every now and then my friend shouted, "Which way?" I pointed and didn't let him down even if I, because of the sleeting rain, did not recognize all the landmarks. However, the lay of the land and the slopes told me we were right on track. The cow paths had become small rivers flowing downhill. I had always wanted to observe the farm and hills under a hard, severe rain. Here was my chance. However, I didn't like this prime opportunity because of the lightning flashes.

Then I saw the fence, the gate, my truck, and horse trailer. We passed through the gate--Everett's gate--and jumped in the cab. WOW! That took us five minutes. Only another ten minutes and the storm should've passed, give or take the five minutes we remained seated under the bank.

Still seeing no break in the weather, the next challenge was to pull out of Bennett Road, a dirt road that becomes muddy and soft with rain. Going downhill ever so slightly was to our advantage and we drove out, pulling the trailer onto the firm ground of McIntyre Creek Road, leaving the wet cows, wet horses, and rising creek behind us.

When you are outside working ranchland, you are in all kinds of weather. In the winter you battle snow. And most farmers and ranchers drive four-wheel drive vehicles which are a help and insurance. My red farm truck is a 2-wheel drive and I have no problems going where I want to go and other places I don't need to go. I try to take care of the land by making no more trails than necessary. Driving home was easy with the windshield wipers going--woosh, woosh, woosh, woosh. By the time we got home and I turned the key, the storm had stopped--twenty-plus minutes later. Now that is timing!

A great weather forecast out on the range under unfavorable conditions. But the crazy thing for me was to use your cell phone to make a call from nowhere to somewhere, where they watch the computer screen and can then predict for you a storm's behavior.

Maybe I should invest in a cell phone. Or take the one with me we have at home! But you know what? I just don't like to be "connected" all the time.

July 1, 1998

Hope

"You can't win them all, all of the time. However, this time I think we won."

Last Friday, early afternoon, as I drove up to the farm I met Dean and Paul on their way to get the second combine. Dean had already taken his, a John Deere, to the field and parked it. I quickly jumped into the pickup and rode with them to go and get a truck to haul the wheat to the Co-op. Everybody was busy now trying to get organized. It takes a lot of driving and shuttling people to get everybody in the right place with the right equipment.

After a little while, Paul brought his combine, a Gleaner, into the field and before long, the two combines were chasing each other around and around the field, harvesting the wheat while trucks were standing on the side waiting to be filled. At one time, we had five trucks going and as we met each other on the road to and from the Co-op, we waved or gave each other the thumbs-up sign--we were getting the wheat harvested.

Paul had joined us so we would get the crop in as there was some talk of rain for Saturday evening. All you need is a breakdown and you lose your wheat. I rode a few rounds with Paul in the air-conditioned cab of his combine and I watched how he carefully raised or lowered the header. I watched how the wheat was drawn into the machine with the big roller--around and around it went.

Soon, we had one truck full of wheat and Jeff took off for the Co-op. A little later I followed with the other truck, an old Chevy. I like that truck even though Bud says it is the worst truck of the bunch. He calls it "Grandma" because it is so slow. Well, it may be slow, but it surely growled and pulled itself over the hills. True, I had to shift all the way back to second gear and at one time I thought, "Do I need the first gear?" But it did climb the hills. Downhill top speed was a bit scary--40 miles per hour.

What was interesting to watch was the behavior of other drivers who had to stay behind me on the double yellow line or when impatient, passed me. In the country, people were patient--waved and cheered you on. In town, it was a totally different story. People were impatient. At stoplights, they passed quickly as it took me some time with the loaded truck to go through all the gears--growling up to some speed. I hoped those people realized I was hauling their bread. I don't think they cared. Being Friday afternoon, they were only thinking of the weekend and how to stay cool. The fact that there were a lot of extra people in town with the Country Stampede did not help things either. This time I was glad there were stoplights at each major crossing.

At the Co-op, I weighed in, unloaded, and weighed the empty truck. After they handed me the ticket, I quickly drove back to get another load.

With the two combines in the field, the trucks kept rolling. It was a good feeling to be involved with the harvest. With as many different trucks and drivers hauling wheat, at one time, I thought Bud had made a very quick trip because he was back already, now waiting in the background. On second look, I realized it could not be Bud because the driver wore earrings. It was Connie. She had come to help us too. Later that afternoon, I passed her coming in with a load of wheat. We waved.

As I drove back, I listened to the steady pulsing of the old engine and I thought of how many loads of wheat it had hauled. A few years ago, it had hauled five-dollar per bushel wheat--this year it might only be half of that. On top of the wheat price being low already, I noticed when I quickly glanced at my ticket, we were being docked quite a lot. Weight was 55 pounds, moisture was okay, ten, but grade was hurting us.

Well, it is like Paul said, "A wheat crop is like the other grain crops--a farmer takes whatever they'll give him for it." I know that a pile of wheat in my garage is no good to me either, so we take it to the Co-op and hope for the best. Remember, Bud's dog name is

"Hope." I talked about that last week. So, we keep hoping that the price of wheat will go up.

I had to chuckle when I overheard some discussion about which of the two harvesters had done the best--the John Deere or the newer Gleaner. It had been interesting to watch them go around and around behind each other as the afternoon wore on and we went from one field to the other, harvesting all 50 acres by evening.

They both did well--they harvested the 1998 wheat crop. And when it rained Saturday evening, I quietly thanked Paul and Connie again for helping. We won that one--we got the wheat off and delivered to the Co-op before it rained.

August 10, 1988

Baba's Farm

"The hills on Baba's farm are all hooked together."

It was early evening and we decided to take one more walk over the farm so that my daughter and two grandchildren, who were staying with us, could say good-bye to Kansas before they would drive back to Ohio the next day.

Joseph, the four-year-old, really wanted to go to the farm and when we said, "Let's go," a happy smile came over his face. He knew the experience. When he was little, he had already gone with us. First sitting in the backpack, later walking, and when he got tired, being carried on someone's shoulders.

Joni, his one-year-old sister, picked up on the excitement but really did not know what was involved. However, she too was ready to go. Being carried in the backpack with her sturdy little legs sticking out and kicking, she could not care less. It was all fun to her. When too tired, she would lay her head forward and pleasantly doze off.

It was a pleasant evening to walk; the day had been warm but not hot, being the early summer. I remember the fragrance of the hay meadow and the quiet calling of the birds. The sun was still above the hills as we set out on our walk. With the children, including my youngest son talking, there was little chance of surprising the deer, but you never know.

What makes it so much fun to walk the farm is the great diversity of terrain and landscape. The farm is cut by several creeks coming from the hills which make the watershed, with a few springs adding to the water flow. During much of the summer, the creek beds are dry, at least not flowing, with only standing puddles of water to provide drinking places for game. Along the creek's steep banks, woods have grown up with big oak trees providing shelter for squirrels, deer, and someday, hopefully, turkey.

While walking through the partly overgrown wooded lanes of the farm, one could imagine oneself to be somewhere east. It is all woods, but as the walk slowly raises itself above the rich bottomland, the oak trees give way to cedars and sumac, and passing those, you step out on top of the hills, the Flint Hills.

It is the hills of which Joseph, the four-year-old, says, "They are all hooked together." Here is an example of how clearly a child can see things and then say it in a simple way. "The hills on Baba's farm are all hooked together."

He was comparing the gently undulating hills along the horizon with a long freight train. Joseph loves trains. Don't ask me why, but he has a passion for them and he knows their make-up--many, many wagons, slightly different yet similar, slowly rolling by. He recognized a similar rolling in the gentle rise and fall of the hills, hooked together in a long straight line along the horizon.

As we walked, Joseph's sturdy legs got tired and he asked if I would carry him. My youngest son had carried him already part of the way, but a sturdy four-year-old is a heavy load for an eleven-year-old. With a swing, I placed Joseph on my shoulders which put him way above all of us. He liked that vantage point and I think he enjoyed the closeness. Sitting on Baba's shoulders is a good feeling.

When the sun had set, we turned around and walked back, following the path through the trees again. As we passed a clearing with a view to the now far hills, we watched the moon rise above them. The birds had settled down and in the distance, the whippoorwill beat out its call--whippoorwill, whippoorwill, whippoorwill.

By now all of us were talking softly, happily walking back along the trail. Joni had laid her head down on her mother's shoulder and Joseph was whispering in my ear. It was then, I think, that he said in a dreamy voice, "The hills on Baba's farm are all hooked together."

Yes, Joseph, the hills are all hooked together on the Flint Hills. That's what makes the Flint Hills special. It is like a big choo, choo, choo train.

August 25, 2004

The Old Saying

Only old farmers may remember having heard the saying "The best fertilizers for the land are the farmer's footsteps." I came across that saying many years ago and when I thought about it, I agreed. But it was the old agriculture that benefitted from it when part of the old way of farming was the smaller farm. A farm small enough where one could walk all around.

Once, I walked our farm with a farmer friend, actually a rancher, and after the walk which crossed fields, creeks, and steep banks, he swore he would never walk the farm with me again. The next time he would be riding a horse. I have often chuckled about that. The worst part was I learned later that he wore cowboy boots.

But I think we are, since yesterday, somewhat even--if getting even is what farm life is all about. Yesterday, I finished clipping the weeds on 50 acres of new brome using the old 20-10 John Deere tractor with the five-foot bush-hog. The old seat shook me like a set of poker dice shaken in a tumbler. And not only that, the steady rumbling of the engine also impaired my hearing, even though I used earplugs.

Going around and around from ten in the morning until after six without stopping, not even for a quick bite, was bone breaking. When crossing the rows at the headlands, the front-end loader bounced and bounced. No springs there--hardly any springs under the seat. And my boney butt paid the price. When I finally quit on the last turn around and had finished the last of the three fields, I felt elated. This job was done and the field looked a heck of a lot cleaner with the weeds knocked down and clipped.

But I also had learned a lot and that is why the old saying "The best fertilizers for the land are the farmer's footsteps" kept popping up. Because as I made circles around the field, swathing every five feet, I looked down and took notice of the stand of brome. Now I have a good picture of my stand of bromegrass. It has been a struggle to get it established.

Two years ago, I had the brome drilled no-till and got a poor stand. Last fall, I overseeded and the two sowings have produced a fair stand with a few places where seed still did not come up. I do not understand it. But talking to other people, I learned that when the first wheat was sown no-till, the kernel would be embedded in the old straw and would not be in soil contact and hence would not germinate or perhaps germinate but could not sustain itself. I now believe that the same principle, or curse, was on my seeding.

Of course, we learn but sometimes you wish other people had learned for you and could tell you that, in some places, the old wheat straw and chaff is too thick and that the tiny brome seed won't make it. It is expensive to learn, but such is life!

With the second seeding last fall, we filled in open spots and as brome will do, it spread its roots and filled in. I noticed all that as I drove around the field. After a while, with my circles getting smaller, it was a good opportunity to check my stand of brome.

Where the old, old farmer walked, I rode and my scale of fields was bigger than the old, old farmer's. However, I also realized that my purring around and bouncing along the headlands was in today's large-scale farmer's eye a joke or an old-fashioned way of doing things. Because to see what I saw, he or she could use the G.I.S. system. He or she could follow or observe while driving their big air-conditioned, many horsepower tractors. Theirs is an office on wheels high off the ground with a soft cushion, no bone-rattling experience--no deaf ears. While going around the field with wide clips, he or she could listen to the markets, make decisions, or phone-in orders. No longer could they listen to this program on KKSU, but that is another story, a sad story that we try to forget.

I did have natural air conditioning on my tractor, sweat which turned cool each time I came to the north end of the field and turned south again. However, as I went around and around, the small, open tractor allowed me to look closely underneath and to the sides as I went along. I was still close to my land and I saw what I needed to see.

Just like you, I get *Kansas Farmer* magazine and I read it faithfully. Recently, I read "Precision Promises Profit" and other articles on large-scale and modern farm practices. I looked at the equipment and tractor advertisements for smooth rides and chuckled when I read my personalized Firestone ad, "Gustaaf van der Hoeven, find your nearest Firestone Dealer." In the picture, everybody is jumping around the combine. They look like Spiderman, fixing the flat in no time. Where were they the day I pumped my tractor tire up?

The more I read and look at that magazine, the more I am aware that our farm is no longer a farm--it is land we love. Yes, we grow a crop--hay. And yes, we run cows on the rangeland. But it will never be a farm again. Yet, we will keep walking it. And as my son so wisely said, "It is not what you make; it's what you don't spend."

September 14, 2005

The Deer Stand

Several years ago, I planted the small walnut trees. In those sixteen years, they have grown. The first years, they did not grow much. I hoed and mowed around them to give them a chance. And as I worked around them, I remembered planting them.

It had started to snow; the ground was wet. Not an ideal situation to plant. However, I wanted to finish the job, so I kept going. It was only a small corner of a larger field, about one acre. But to drive into that one acre you had to cross a drainage ditch which often was not passable after a rain. So, after the last crop of alfalfa, I decided to plant trees there. Black walnuts would be just perfect for the deep, black soil along the creek and that is just what I planted.

As the trees grew, the deer found them and started rubbing. It annoyed me but there was little I could do besides shoot the deer. Some of my friends gladly volunteered. The deer, however, have been smart. Even though there have always been plenty of footprints beside the rubs, as well as the fact that trails which I keep open come together there, I don't think that anyone ever shot a buck or doe in that corner of the field. But that may change.

There is a windmill left from earlier years, before my time, standing at the base of what is now a big oak tree. The steel frame is still good, and I have always wanted to build a small platform about a third or halfway up. The little step ladder would be a convenient way to climb up to it. Sitting there would be more comfortable than sitting against a tree trunk fifteen or eighteen feet up on a small deer stand seat, lashed to the tree. However, I have not yet constructed the platform, so when a young friend of mine asked if he could hunt on the property and I said, "Yes," he had to come up with his own contraption. A seat attached to a sturdy tree to which he must climb up whenever he wants to use his deer stand.

To get an idea of what activity was on the trail below the stand, he placed a camera with an electronic eye below the stand. As most things which have to do with hunting are camouflaged, at first I did not see the small contraption, but my eyes followed the flattened weeds from tramping to the trees and then I saw it. So with everything in place, I now wonder if the deer will have a chance. It being bow hunters who hunt them, I expect they have.

Yet I feel that with so much preparation, things are not totally equal and I got an idea which, with the help of a friend, I would accomplish. The camera eye intrigues me, the all-seeing evil eye. I heard it click when I passed it on the trail. So I assumed a photo was taken. What was done once can be done again and a nice collection of photos may be what my friend needs. Maybe that is all he gets if the live deer are smart enough to stay away.

In the garage, I have kept a deer skull with a good set of antlers. I found it a few years ago when I came across the remains, a full skeleton, lying bleached against some shrubs along a gully on the farm. I took the skull home. But now was the time to use it. Somehow, I had to rig up a scenario where this skull with antlers dances in front of the evil eye of the camera. It was not too difficult once my imagination and creativity got loose.

I want this hungry skull to browse along the edge of the woods so the camera would see the movement and "click"--record it. It would not be too difficult to do. The other photo would be of my accomplice in the deception and me both wearing antlers and sparring in front of the camera. That could make a second "click." My friend has a huge moose skull and doesn't need much encouragement to wear the moose antlers and come banging through the underbrush--there is a moose loose.

Of course, all this must be done before the bow hunting season opens because, with all this foolhardiness, I am sure of one thing: that my hunting friend will feel obliged to let his arrows go and have equal fun. And since I will be the buck, I cannot even throw rocks at him. All I can do is rub my antlers and jump away. No way will I let him rip my flesh with flying arrows, nor will I hear his laughter. What I would like to hear is what he says when he looks at the photo and sees this slowly browsing, floating deer's skull with a nice rack coming into focus.

"What the heck is that? There is no meat on that buck--none!"

October 14, 1992

Sunday Ride

Slowly, it will all turn back to the soil.

Sometimes a weekend is just too short. This past weekend, for instance, I did not know how to fit it all in. It started when my son said, "Dad, this weekend we should go ride on the farm."

When I tentatively responded with, "We'll see," his mother got mad at me and gave me a fast lecture on how I had my priorities mixed up. She reminded me how excited he had been when both of us rode on the farm; actually, we had taken the cows off the pasture and afterward had stopped by the farm pond where his pony decided to take a roll in the cool water with him still in the saddle. Only my warning and his quick response had stopped his pony from going down all the way. That was just a few years ago.

The memory of that ride made me realize how many weekends we all had been busy with other things. Often, we were busy with horses, but not riding them on the farm. I realized, in time, the mistake I was about to make, so I said to my son, "Yes. Let's go ride on the farm. The weather will be great and there is fall color to enjoy."

The Saturday was already spoken for with things to do so we decided on early Sunday morning. I slapped my son's hand and told him how glad I was that we were going. I really was. Once decided, I looked forward to it because I have missed the riding. Oh, yes. We check the horses nearly every day and give them a treat but that is all the time we have. We think. And the more I think about it, the more I know my wife was right. My priorities were mixed up. But I have to work to keep these horses and the rest of the family fed!

When we got to the farm, we each grabbed a halter and set out to find the horses which, of course, were on the other side of the hills. They neighed when we got close. We haltered them and walked them back to the truck where we tied them to the rail. After grooming them thoroughly and bringing the shine back to their coats, we saddled them up. The pony took it all in stride as if he was ridden daily. My horse, with his Arab blood, is a bit more temperamental. However, it is an honest horse. Just to be sure though, I lunged the horse to make sure some of his energy was spent. It did not help much because when I mounted him, he had to buck. But it was all show and once I talked to him and set course for the alfalfa field, the horses stepped together with eager strides.

Horses belong to the west and I enjoy riding. I don't mind my horse being spooky because I know him. We rode the hills and checked the cattle. I showed my son the quality of the range and explained grazing pressure and selected burning. Last year, we burned the north half of the range, hoping that the cattle would graze the sweet, new grasses and give the south half a break. Looking over the hills, it worked. Next spring, we will have a good burn on the south half where there is plenty of fuel left standing.

We rode along the edge of the hills and looked out over the bottomland where the swathers were busy mowing the last cut of alfalfa. My son noticed how beautiful the valley looked. He remarked about the swaths of green alfalfa hay lined out across the field. On the way back, we tried to count the cows and grown calves. It is easier for me to count the bull. There is only one. The calves kept pushing forward, poking their heads up, over, and under. I gave up!

As we rode on, a prairie chicken exploded from cover and caused my son's horse to sidestep. I laughed as my horse simply ignored it. When we came down over the hill, I saw a vulture rise and went to look--a dead cow. I told Martien, "There lies a thousand dollars." The coyote and vultures will take care of the cow. Someday I will find some bleached bones scattered about. The mice will gnaw on them and slowly it will all turn back to the soil.

October 13, 1999

Good Fences Make Good Neighbors

There is an old saying, "Good fences make good neighbors."

Oh, it was a beautiful sight, but each time the cow moved I could hear the pods snap and the beans pop out!

It is this time of the year that if you run cattle, things tend to go wrong. They seem to get it in their head that grass across the fence is greener, especially if the grass is on the neighbor's property.

A week ago, Thursday evening, we went out to check the horses and walk our property. Even if I don't count the cattle grazing on the southeast part, I surely like to see them while I am standing on the hill of the horse pasture. My wife tends to spot them first. From that distance, looking across the bottomland, you can't count them, but you quickly see if they are grazing quietly. This time of the year, it is getting hard to see the roan-colored cows as the grasses have turned to the gorgeous fall colors of red, auburn, and orange. The colors are vivid, especially as the sun goes down.

We walked along the ridge, stopping at our favorite lookout, but did not see anything out of order. Way in the distance on the other side of the valley, we could see the cows slowly walking and grazing away from the big pond. As we walked back along the ridge, the dog decided to go down into the woods and see what he could scare up. After a little while, I heard his bark. And just like with your child who is crying, I knew he found a cow. My wife just thought that he had spotted a squirrel or treed a possum, but the bark was different, more intense and more businesslike.

And yes, as we came down from the hill, I saw his quarry--a feisty Angus bull which did not like to be barked at nor challenged to get out of the soybeans. He was way up to his belly in the beans, tossing his head around as if ready to charge. I walked up to him but quickly realized that on foot I would do more harm than good and that I would only shake more beans loose. So, I just stepped back and looked at him.

He was a nice bull--shiny black. My brother used to say that Angus bulls were very dignified, all in black. I think he saw the clergy in those big bulls. Looking at this strong bull, I saw a pretty picture. A picture of yellow, bleached beans with the coloring trees in the distance, the amber hills with coloring grasses and topping it off, the clear, blue evening sky. The colors were great--black, yellow, amber, and blue. Once again, I should have had my camera, but all I had was a white handkerchief to declare a truce. The bull warily looked at me, tossing dirt and beans while pawing.

By now the dog had stopped barking and stood off to the side, watching me and the bull. "What now, boss?" must have gone through his mind. I praised him for alerting me of the presence of the bull and called him off. It was only with reluctance that he followed me out of the bean field to the pickup truck.

As we left, I closed the gate, not wanting the bull to wander along the road at night. I have often told people of my fear of running into a big, black bull on a pitch-dark night. I once nearly did. It was near Beloit. Only that night, it rained cats and dogs and a man could barely see the road. And suddenly, there was this big, black cow walking along the yellow line in the middle of the road. I would have hit it just below the tail. That cow would have made a hell of a nice hood ornament! Like the old ram's head on the Dodge pickup truck.

I kept wondering where the bull in my beans could have come from. I later found a clue when I saw a group of Angus cattle grazing the range with the same color ear tag. Once home, I made some calls and found out "so and so" looked after the cows and "yes, indeed," they had been looking for a bull. Apparently, someone left a gate open and that caused a stampede. It is like the old saying "shit happens," but this time it happened in my beans! But you know, I can still see that black, feisty bull standing there in my ripening bean field--a pretty picture it was. Black on yellow--blue sky.

Saturday evening, we again walked the ridge--guess who we met grazing on top this time. Yes, you're right. The bull. He had busted the rail gate to get to my salt lick. I think I'll buy some heifers of my own and let them graze with him. He ought to like that! It would be one way to start my own herd!

November 18, 1998

Fishing

"Baba, we need more worms."

This past Sunday, Matthew and Ryan, two little van der Hoeven's, their parents, and the dogs drove up the driveway around 11:30 in the morning. It promised to be another beautiful day after a beautiful Saturday.

And by the way, what a Saturday it was. Even though it was gorgeous weather, I spent most of my time watching the Kansas State vs. Nebraska football game on TV. Having been inside that day, I had one more reason to get outside on Sunday, and besides, the boys wanted to come out and fish in the big farm pond. They had called early that morning to say that they were coming.

It could not have been a more beautiful day for mid-November. There was little wind and what breeze there was, was gentle and pleasant. The drive out to the farm showed the Kansas Flint Hills in all their beauty. The grasses have colored the hills brown-gray. Depending on what time of the day it is, the landscape has a violet-pink to purple hue. Cattle were grazing on the hills; their silhouette looked like cut-outs.

When we got to the farm, we first took a hike to see how the horses were doing. They have their winter coats and still seem to enjoy the free run of the hills. Once the pond freezes over, they have to come home as there won't be any water for them to drink.

As we walked, the two dogs, a Labrador and a Golden Retriever, roamed around us, kicking leaves and sniffing. The farm was paradise for them compared to their large pen at home. Several times, we had to cross water which was for the five-year-old, Ryan, an obstacle. But with a little coaxing, he walked the plank which crosses the stream while his mother and I held his hand.

The farm holds a certain mystique for them. For them, it is still a very big place. They are only five and seven years old. They talk about the farm, but they don't really know it. They have not explored it yet as the older grandchildren have. But they are learning. While walking the trail, Ryan asked, "Where is the farm, Baba?"

I told him, "You are walking on it."

His big eyes looked up at me as if he was thinking, "That is a thought."

After checking the horses, we went back to the car to get the fishing poles. Each of the boys carried his own and then came the rough part; they had to walk the two plowed fields. Their short legs and little feet had to step over and on big clods. But they made it and after several times asking, "Where is the pond?" they could see the water in front of them.

Not being much of a fisherman myself, I let my son handle the chore of readying the tackle. However, my job was to look for worms. It took a little while to coax a fat worm from the cold soil. They had already crawled down to get away from the frost. But I found one and now with a worm, the boys were in business. I was surprised to see how well they could cast the lure and bobber to the middle of the pond. They were persistent and kept working at it.

I sat down on the bank and watched the young family have fun. Ian has the patience to show the boys how it is done. I watched the gentle breeze play ripples over the surface of the pond and saw the cows looking down at us from the hills. Vicky took the dogs for a walk, one of which growled at the cows as if she felt that her boss needed protection from those unfamiliar, very big dogs--the cows.

After a little while, Ryan shouted that he had a fish. We had some earlier false alarms but this time it was for real. He proudly reeled his line in and lo and behold, he had caught a six-inch catfish. Vicky, his mother, shot a picture for posterity. Little guy--little fish--big pond. The tiny fish was released back into the water for another day. The boys kept fishing but soon I heard, "Baba, we need more worms."

I grabbed the spade. This time I knew where to look. I turned over a dry, big cow pile and yes, there was a nice worm ready to become fish bait.

The scene--father, two sons fishing, mother, and two dogs--was the perfect picture. I saw the cows come down the hill and graze slowly toward us and to the pond. At times, they looked up and gazed at us the way only cows can do--big, soft eyes and big, slapping tongues.

When the sun went lower, we decided to call it quits. Matthew will catch his fish next time. The two dogs took one more swim, came out, stood next to us, and shook themselves--thank you. The cows looked at us as we packed up and left for home.

A great day! There can't be that many more before winter is here.

November 5, 1997

A Windy Weekend

I never sensed this past weekend that the wind was blowing strong--I accepted it as just a Kansas day.

I didn't even think about it but they were Kansas days all right, with all the wind, blowing dust, knocking leaves off the trees, and making waves on the lake. Tuttle Lake looked rough with waves splashing onto the shoreline and nasty white caps topping rolling waves.

It was a great weekend. I walked the farm and enjoyed the shelter from the trees once I stepped into the woods and followed the trail to the presently dry beaver pond. As I walked, I listened to the wind roar through the crowns of the trees but did not mind. I kicked the fallen leaves and picked up some of the acorns--Bur and Chinquapin oak. I put them in my pocket to scatter them somewhere else where some oaks would look good. If only one or two of the acorns germinate, that would be all right. On my next walk, I'll pick up another pocketful and scatter some more. It is the easiest way to plant a woodlot. And if mice or deer want to munch on them, so be it. It is all part of nature.

As I stepped out onto the small meadow, a hawk lifted off and was taken by the wind. It righted itself and swooped low, then rose again while making its high-pitched call. I love their call. Closer to the ground, in the protection of the shrubs and out of the wind, chickadees seemed to follow me with the chickadee questions, "What are you doing here?--What are you doing here?"

When I passed the old beaver dam, I noticed a large turkey foot imprint in the mud. It was big and you could easily see why the tall bluestem grass can be recognized by its three seed heads in the shape of a turkey foot. It was clearly marked in the soft mud. I am glad the turkeys are back or are just visiting to take their pick of the acorns and other seeds they like. This summer there were plenty of grasshoppers to stay fat on, so maybe they have been around, but I just had not noticed.

Kneeling down to take a closer look at the turkey foot imprint, I noticed the perfect yellow leaf of the cottonwood. The wind must have torn it off, but I expect that the yellow leaf knew it was time to let loose and when the next strong gust of wind came by it probably said to its neighbor leaves still holding on, "Here I go--whooopeeeee!" The other leaves, a bit more timid perhaps, just could not let go--maybe they were not perfect yellow yet. But sooner or later they too will let loose and will drift down.

The leaf I looked at must have had a wild ride on the bucking wind. As I looked up, I saw other leaves ride the wind and thought I could hear them shout to each other, "Whooopeeeeeee, what a ride!"

Once they fell to the ground, the wind pushed them further along, blowing them into heaps where they will make a mulch, protecting plants from frost. And as they decompose, nutrients will return to the soil to once more be taken up by the tree roots to form new leaves in another season.

It is the everlasting cycle of birth and death of which we too are a part. Just as the leaves have a function, photosynthesis, each of us has our job to do between the beginning and the end. I wish that we too could be assured of such a glorious end as the yellow leaves floating, bucking, and dancing on the wind.

Kansa, the dog, was looking at me. The wind ruffled his coat but he too loved it. Only, with the wind shaking and rattling everything, sending leaves rolling along the ground, he could not make up his mind what to chase first.

Before I went home, I finally decided to cut some firewood. The old chain saw started on the second pull for which I said, "Thank you." Yes, I did. Then I quickly cut up wood from a downed tree. The chain was sharp and the cutting was easy.

While loading the wood, the sun was going down and the evening sky was showing color. The color is different now that summer is gone. The sky already looks colder, even though it doesn't really feel colder yet, especially not if you just cut firewood.

Driving home I watched the hills. At times, they seemed purple as the pale, orange sun was getting low, slipping away between the clouds--going--going--gone!

December 19, 2012

Walking with Amanda

Sometimes I walk on the farm and the person I am walking with is not there. Yet, as I walk, I talk and listen. It is not because I am losing it, but because I am thinking of the person who is not with me.

The other day, there was an early morning mist and the woods were really alive. I walked with Amanda, my granddaughter, but she was not there. I pointed out to her all the things which are on the farm, things like picking up the big feather of a turkey or the smaller feather from a Blue Jay which is so much more intricate. We both looked with only one pair of eyes.

There were the big imprints at the creek crossing of a deer and the very tiny ones from the fawn. We compared them. I showed her the direction they had traveled. I told her about how the bucks scrape the velvet off their new antler growth to mark their territory. When we passed a tree where the bark had been rubbed off, I pointed it out. "See, this is where he stood rubbing," I said. This was a big buck as the bark has been taken off quite high. Amanda was impressed.

As we walked along the prairie, I pointed to another tree where you could see low rubbing and said, "Now that was a very young buck. He did not reach up very high." I tried to make her see the buck scraping away and then walking off with strips of velvet hanging down like Christmas ornaments from his antlers. She laughed.

She easily kept up with me. Happily walking beside me--not being there. We passed a patch of bittersweet. The vines had climbed and twisted around the shrubs and into a burned, hollow stump. I picked a small twig with the orange bittersweet fruit and we both took a closer look. Her interest was brief, but she looked as I pointed it out.

On we walked and I talked, keeping her attention. When we reached the old hollow tree in the woods, I told her the story of the tiny gnome who lived in the tree with his family and how they had rooms and stairs going up high into the tree. I pointed out the chimney. There used to be a branch there when this tree was small, it broke off. The gnomes used that knothole as a chimney. Amanda laughed and thought it was interesting to tell stories, make-believe stories. She agreed storytelling made the woods more intriguing.

We continued following the trail the deer walked into the woods. She kept up with me as we next climbed up the hills. We walked up the long valley past growing Bur oak trees. As we got higher, we stopped where there are bones scattered from Flash, my horse who was struck by lightning and killed several years ago. Each time I pass, I take a moment and pile the bones back into a neat pile. With the two of us, it did not take long. As we picked up the sun-bleached bones, I explained which bones were which and where they had fitted in Flash's skeleton. No longer could I put a whole horse together but there was enough for Amanda to get the idea.

We walked further up the hill and once on top, we looked around in a wide 360-degree circle. On top of the hills, the wind blows and thoughts tend to blow away--away. Walking further on top of the hills, we came across a remaining buffalo wallow with its hard-caked bottom where hoof prints could still be seen. I told her about the wallows and then we pretended that the buffaloes were all around us, big beasts grunting and snorting in their shaggy fur coats which helped them to withstand the winter storms and snow. As I talked, she saw the buffaloes but was not afraid as we stood close together.

Next, on top of the hills, the Flint Hills, my horses came to us. Misty and Poptart came close, expecting a handout of grain. They were not disappointed. I showed Amanda how to hold her hand flat so that the horse would not pinch or bite. Their soft lips and tongues wetted our hands and tickled. I told her to talk to the horses so they could hear her voice. It was not necessary to speak loudly; soft, gentle talk was enough. She watched the horses' ears as they listened to her and ate the grain out of our hands.

Giving the horses a pat goodbye, we walked on together down the rocky hills to the dry stream. There we looked for a specially shaped piece of flint, the hard rock for which the Flint Hills are named. Oh, there was so much to share and talk about as we walked together all over the farm looking at things, listening, and pointing things out.

Walking further, we followed the valley until we stopped at the old farmstead. Now, only old stone walls stand. But as we sat on the wall and rested, we could imagine the family living there long ago. At the fire pit, we stacked some wood and then I lit a match, the wood catching on fire and the flames rising. The warmth of the fire felt good. You know, one can be very much together, yet alone, yet together. Thanks for a great walk, Amanda.

August 29, 2012

Farm Trails

Our farm has lots of places where one never goes. It is when I look for a lost cow or newborn calf that I walk down the gullies. You can see into the open gullies where no trees grow. But the older gullies, over the years, have become deep and steep draws that are thickly wooded with big cedars and oaks. You never go down there unless you need to or want to. Over the years, I have wandered down at times in winter when the leaves are off and visibility is easier. I have crawled on my hands and knees to get through the tight places. The low hanging, dead cedar branches can make an effective barricade as can a tangle of shrubs. These off-the-beaten-path places are interesting parts of the farm.

Production-wise, grazing or growing these areas is a waste. No tonnage of hay harvested, nor pounds gained on cattle from those areas. Yet, I don't see the land as wasted. It is part of the farm and it is habitat for wildlife whether it is deer, turkey, bobcat, squirrel, or whatever chooses to live there. Ever since we stopped opening up the woodlands for grazing, underbrush has come back and it seems we have more and more variety of songbirds. I like that as I like to listen to the call of the owls.

To help open these closed and wooded areas, as well as to encourage hikes and trail rides, I started years ago laying out an interesting pattern of trails on the farm. When laying out the trails, I tried to follow old, overgrown farm roads. By carefully studying the land, I could find old paths. I have not yet reclaimed all of them but I'm working on it. There is a section along an old rock wall, halfway up the hill, which ran right through the center of the farm, connecting one house to another. Looking at that old trail, I can imagine people walking or riding it by a full moon on the way home after a visit or helping with the chores.

Those are the footprints of the people before me who farmed our land. Like old Bud used to say, "They grew tall corn on that bottomland." I have never grown corn; we have grown milo, wheat, oats, beans, and alfalfa. Right now, it has been in brome for hay for many years. What I really like is a thick stand of grasses--mixed grasses, forbs, and clover. A hay meadow like that, made up of cool-season grasses, needs more rain than we get. Keeping a healthy stand on the range is already a challenge. But with the rain we got this past weekend, it is much easier. I stepped outside for a while and just got wet, nice wet, and said, "Thank you--Thank you."

I had started mowing the trails before the rain came. It is amazing how tall the weeds can grow through the summer. But once mowed, the paths reestablish the existing trails. Now with the trails opened up, the wildlife tends to follow. I have observed deer tracks in the snow along the trails. When it's muddy you can clearly see the prints in the mud, the big ones for the buck and the very tiny ones from the fawn.

While mowing the trails last Thursday, I came across a leg and spine, part of a big deer. It still looked fresh, although the meat was gone. Coyotes might have pulled it away from the killing spot or was it killed by a hunter out of season? I doubt if he or she would have taken more of the carcass for meat. Wanton killings I don't believe in. If you hunt it, eat it.

On the farm, while building new internal fences, I consciously moved the new fence several feet so that I could have the trails to walk and ride on that I wanted. Doing that opens up the farm for a different experience than plowing, mowing, or working the field. Walking is a slower movement. You see, hear, and feel more. The trails through the small woodlots interconnect the fields.

While walking, I noticed that we have less wildflowers this fall. Every now and then, I came across a turkey feather. Once I had a bundle, I stuck them together into a fence post behind the wire. Butterflies seemed to follow me and there was fall color about. Every now and then, I walked into a spider's web woven across the trail. I apologized for breaking it up, "Sorry." The trail was busy. Two big, fat tom turkeys walked away from me.

Our farmland is not a square 435-acre wheat field, beautiful as that is. Our farm has steep topography, rough land, and woods as well as creeks and prairie meadows. It has tremendous diversity and to enjoy it more, I open it up with trails. And mowing them twice a year to clear them is great fun!

Landscape

Precision farming.

January 8, 1997

Knitting a Yard

It was dark outside. Looking from the window, I could see the lights of Manhattan in the valley below us. There were still some red and colored Christmas decorations celebrating the extended holiday season. But each evening, more of the lights have been taken down and the town starts to look its old self again with the urban lights ever-expanding toward the west and on into the hills surrounding Manhattan. It is amazing how this university town has grown over the last 23 years. You really notice it during the dark winter evenings when you watch the lights.

I turned around and laid some more wood on the open fire. Kansa, the dog, opened his eyes and looked at me as if saying, "Life could not be better." I picked up my book and continued reading as my wife was knitting, knitting another woolen sweater. Every now and then, she asked me what I thought of how the pattern was developing. The sweater she is knitting has a Norwegian pattern, but she was modifying it as she was working and wanted to be sure that it was just right. She has knitted many sweaters for me, for our children when they were small, when they were teenagers, and now for some of their spouses. I only wear the warm homemade woolen sweaters she has knitted, and I have a wide selection to choose from. There are some I cannot wear to work anymore, at least that is what Anneke says.

Knitting with warm wool is a winter job. You cannot imagine doing anything like that in the summer. So, with the added logs on the fire, I watched her needles fly one over, under, through--one over, under, through--or however she does it. I watched the two long columned woolen threads become a pattern, the pattern become the back, a broad back for my daughter's husband. It had all started with the careful selection of colors and the next step was to wind the strands of wool into the ball she was now knitting from--one over, under, through.

As I watched her hands work the thread into a pattern and a back for the sweater, a thought came to my mind. Somewhere there had been grass growing. It had rained and the sun had been shining and the grass grew. A flock of sheep had grazed the grass. At shearing time, the sheep had been driven to the shearing shed. Their fleece was taken and the sheep returned to pasture or range. It could have been Wyoming, perhaps somewhere in Australia.

The wool then went through all the processes of cleaning, sorting, spinning, and coloring, and it became yarn. A yarn by itself is nothing. You can tie a package with it, but you should not pull too hard. It is when wool is knitted into something that the strand of wool gets its meaning. It becomes something: in this case, a sweater and it can then even express love.

It is the same with words. Words by themselves mean very little. Words forming sentences can tell a story. They can create poetry or, combined with music, words can become songs. It is the same with a meal. Rice, potatoes, vegetables such as carrots, peas, broccoli, are all that--vegetables. But when combined and prepared, they can become a meal and depending on the occasion, a dinner celebrating an event.

It is the same in landscaping. Grass, trees, shrubs, ground covers, flowers, ornamental grasses each by themselves are just objects--like different threads of wool, different words, different vegetables. They mean nothing. Oh, they can be beautiful while standing in the field or nursery row. Just like a strand of wool can be of beautiful color and quality. But it is waiting to be made into something. The trees, shrubs, flowers are waiting to be made into something--a home or farmstead landscape. Creating privacy and turning a space into a place, making a sweater from a thread, creating a story from words, creating a landscape.

Knitting a beautiful pattern in a sweater is an intriguing art. You see it come to its full form as the needles click and click and hands and fingers direct. A home landscape is the same. You take trees, shrubs, flowers, and a little bit of lawn and whatever else it takes--a pond, a rock garden--and you create a place. Your place. A place to relax, to sit and enjoy nature, songbirds, and the rhythm of the seasons--even the rain, the snow, and the ice.

As I threw another log on the fire, Anneke held up the knitting. "How do you like it?" she asked. I liked it. It was a nice pattern and will look good on our son-in-law.

Why don't you start "knitting" a home landscape now while the winter evenings are still long? There is a lot of reading material, even videotapes, in the Extension office and library to help you with landscaping your home grounds. Make a good start with planning on paper and then keep "knitting" until you have a landscape that fits you like a warm or "cool" sweater.

Come to think of it, even 1997 is like a single strand of wool to be made--day by day by day--into a year. I am "knitting" for a happy year and I am wishing you the same.

January 12, 1994

Hoar Frost

I should have crossed the fence and taken a hike. Driving out of Manhattan yesterday morning, I left a gray town under low overcast skies. The streets and parking lots were markedly busier than they had been for the last three weeks with the returning students. But leaving town, it did not bother me and besides, students are the life of the town.

As I drove out, my car climbed the steady incline of 177 South which takes you to the top of the Flint Hills. I have driven the road many, many times but each time coming home or leaving town, I experience the difference in the landscape. Coming home you dip or drive down into Manhattan as the town, or at least most of it, lies low near the river. That is where the town was started. At night you see the city lights below you as you drive down into town.

But when leaving Manhattan, you climb up onto the Flint Hills. The change is quite dramatic and once on top, you can see the vast expanse of gently rolling hills ever-changing throughout the seasons. Yesterday, it was different again--the grasses, the trees, the fence lines were all covered white with thick, hoar frost. The view over the hills and into the valley was no more than half a mile--or even less. The hills became the sky. There was no separation. Everything was gray except for the white frost which clung to everything--beautiful. Grasses were standing erect, each a solitary ice statue shaped to the contours of the grass or silent weed. Shrubs and trees showed their dark branches outlined in white frost.

If I had not had an appointment, I would have stopped the car, jumped over the fence, and walked the prairie. You wonder why you don't do that. You drive by seeing it, but you don't partake of it by moving into and becoming absorbed by the silence. I know from other times what it feels like to walk through such an ephemeral landscape. With no wind blowing and limited visibility, it does not take long to be in a world apart, the silent world of winter. Along the wooded edge, the trusting and inquisitive chickadees follow you with their cheerful chatting.

As I drove on and turned east on I-70 to Topeka, the winter landscape covered all of the Flint Hills. I believe it was near McFarland that the white frost thinned out and visibility became near normal for a gray, overcast day. Driving on, my thoughts turned away from the fairy tale landscape I had viewed earlier.

Driving home in the afternoon, I noticed as I came closer to Manhattan that the white winter landscape was still there as the sun had not come out and the temperature had stayed around freezing. Nothing had changed from early morning; nothing had melted. Grasses, weeds, shrubs, trees, and fences all still were covered with the frost. The grasses and weeds were ochre-brown with white; shrubs and the trees a darker gray and black. I watched hawks sitting motionless on tree branches, fence posts, and even road signs. Their dark body feathers on folded wings, marking them clearly against the white, frosted branches of the trees.

I now was hurrying home and again I did not stop, cross the fence, and take a walk in this winter landscape. Turning off I-70 and following 177 into Manhattan, it made me wonder yet again why we do or do not do things. As Robert Frost said, "Miles to go and promises to keep." But doggone it, would it not have been fun to pull off the road, turn off the humming engine of the car, step out, take a deep breath, jump the fence, and leave the world behind to get lost in the special world of a winter landscape, which after a half-mile of walking would have swallowed me up in the soft, gray mist?

Next time!

May 7, 2004

Robert Sudlow

It always amazes me how a painter can paint on a relatively small canvas or board--say three feet by five feet--a vast landscape and sky which encompass hundreds of acres, if not more. The same holds true for a landscape photographer.

The expanse of the landscape held or captured on canvas would make some great farmsteads if all that land belonged to the one farm. It seems that farms give the painter greater satisfaction to paint as there is more texture, form, and color to be creative with. Especially when seen from a higher vantage point.

Imagine yourself standing on top of Coronado Heights, just a few miles northwest of Lindsborg. When you climb up onto the viewing platform of the stone castle, you look south and east over the quilted landscape of agriculture. There are fields of different sizes and shapes--some rectangle, some square. They have different colors this time of the year. The wheat fields are a deep wheat green. There are plowed fields that show the deep brown earth tones. There are the hay meadows. Many fields are separated by the greening hedgerows or windbreaks. Looking at a photograph or a painting, it would hold thousands of acres. Quite a nice farm!

If you turn around and look west or northwest, the scene is of native prairie. Where the agricultural view was flat land, this is hilly land not suitable for plowing. The land is for grazing and this time of the year the native grasses slowly turn gray-green. The distance from the top of the lookout is too far to see any of the early colorful prairie flowers. The yellows and the purples all hugging the ground as if still afraid for the cold north winds which only a short while ago blew across. There are no hedgerows across the hills. No deeper green of budding and leafing-out trees. The hilly land is divided by barbed wire fences running up and down the hills.

So turning around, a painter can choose his subject matter and look for inspiration, be it agricultural or native hills. The one element which holds both landscapes down is the sky, and it could be that a painter or photographer suddenly decides to capture a sky and only uses the horizon as a baseline. There may be four-fifths sky and only one-fifth earth. However, I have seen the reverse too--four-fifths earth or farmland, a valley, and only one-fifth sky. It all depends on the reflective mood of the moment and what one wants to capture.

If one would spend a whole day on top of Coronado Heights and observe the changing light fall across the landscape below, one would see many different impressions. This could be captured in one photographic moment or quickly painted on canvas. Whereas the landscape--the arrangement of the land--would not change noticeably, the light fall would give the landscape a continually different look. That would be especially so if there were clouds, storms, and early morning fog. The faraway rain squall slowly moving along and holding a curtain of the falling rain has always caught my eye, especially when one stands high.

It must be very difficult for a painter to capture light. I am not talking about light fall, a ray of light coming down through an open window. I am talking about the atmospheric light which I think we take for granted--day versus night, night versus day. I am referring to the light we move through during the day. Do we ever look at that light like a painter who is trying to capture it on canvas to hold his painting down? To hold the earth, be it organized agriculture fields or our vast, native prairie.

Last week, I slowly walked past the exposition of paintings by Robert Sudlow presently hanging in the Beach Museum here on campus. I plan to go back and stand longer in front of a few of my favorites. *Cold Front,* painted in 1990, caught my eye. It was as if I could feel the cold air, watch the fading, moving light, and hear the crunching snow. I stood still for a long time in front of *Snow Road to Tucker's Farm,* painted in 1989. There were others including *Road to the Sky,* painted in 1991. The paintings I looked at were Kansas scenes with Kansas landscapes and Kansas light fall. Robert Sudlow set up outside, painting valleys from vantage points, braving the weather to capture that fleeting moment on canvas. Valleys that have been plowed up for farmland, for growing crops, make for interesting form, texture, and color.

I am not a painter. I do take photographs. But walking outside and looking--seeing the landscape--makes one appreciate the ability of a good painter to capture what we see in the landscape and bring it home.

February 14, 2003

Landscape Design versus Rocket Science

A few weeks ago, I was asked to write a short article for the Nebraska Nursery and Landscape Association. The article is related to the Nebraska nursery conference, which was held two weeks ago. The article is about landscape design--what it is and what it is not. I made a statement which in hindsight could be regarded as a prediction, looking into the future. However, when I wrote the article, I had no premonition of what would happen to the Columbia space shuttle upon its return to earth. Anyway, this is what I wrote in the short article and what I believe as far as landscape design is concerned: "I hold institutes of learning responsible for how and what we teach."

Not long ago, I overheard someone say, "Landscape design/landscape architecture surely is not rocket science." Hearing that statement, I raised my eyebrows. Perhaps knowing the source, it should not have been too upsetting but the fact that the source is an academic administrator does worry me. True, to put a landscape together, to create a landscape, is different than putting a rocket together. However, both can fail and if the rocket is a manned spacecraft then there is a loss of life.

It was this administrator's statement which reflected on the earlier loss of lives that brought the Challenger to mind, as well as other mishaps. I did not know that a few days later the Columbia would disintegrate on reentry. I still believe a landscape design can fail horribly, but, true, not at the cost of lives. Nevertheless, landscape design/landscape architecture is a science and an art. I have always told students, "Given enough time, I can make you learn the science of plants, soil, etc. However, the art is another matter." If you have no creative, artistic touch or imagination, then designing a landscape in a garden with a deep sense of place may not be possible unless nature takes over.

Landscape design is a process that generally plays out over time. It is based on science and art following a chosen theme which could be derived from the site, culture, or function--among other options. The theme directs the style and use of landscapes, all held together by the carefully selected, adapted plant material. The selection, the right selection, of plant material being trees--large, medium, or small--shrubs, groundcovers, vines, flowers, and ornamental grasses is a science. It becomes art if we select for form, texture, and color.

Designing a landscape does have a certain mystique. Not every horticulturist is a designer and is able to turn a space into a special place. Too many people plant and just decorate, but that is not landscape design. Consciously and subconsciously, good landscape design is therapeutic. It feels good to be in the garden. It provides one with a sense of place.

You can ask yourself, "What is a landscape?" There are many books written about landscape design and nearly all have some statement in regard to the definition of landscape design. I happen to like my own: "A landscape is an outdoor space organized by using plant material." And what about this one which I have used for years: "A landscape is the arrangement of plant material by form, texture, and color in such a way and on such a scale that one can walk through it, that it gives a feeling of privacy, but above all, it gives a feeling of being somewhere." To me, that is what landscape design is all about.

The final question I have is this: how many people will travel to the moon and at what cost? If one compares that with how many people enjoy a garden, the home landscape, a community park, or landscaped corporate grounds, one quickly sees the importance of good landscape design/landscape architecture.

We need good, creative landscape designers and landscape architects who know their plant material. You can develop greater creativity in design, but you have to work at it--read, travel, and look around you. Do not accept mediocrity and above all, do not decorate!

May 10, 1995

Buffalo Wallows

When I saw them, they were slowly walking away from the pond, walking uphill following the cow path made years before. I followed them along the hillside and watched them reach the flat hilltop. By now they had seen me and were spooked. Instead of spreading out and grazing, they started to trot and ran away from me. I slowly followed them, studying the new grass.

We did not burn this year. One weekend we could have, but it rained. The next weekend, the new grass had grown too much and we found we would lose too much early grazing. So, the decision was made not to burn this year. There was plenty of grass for the small herd of cows, but they were still spooky. Now they had moved way beyond me but had stopped and bunched up and were looking at me--a small figure slowly walking toward them.

I looked at them through my binoculars but could not make a count. As I drew closer, they circled around me, all bunched up, and stopped again. This time they all lined up and faced me, a mixed bunch of cows. I suddenly realized why I could not get the right count and kept being one short. It was that black cow with no white which made it blend in with the other black cows. The others were red and roan like I said, a mixed bunch, but all good mothers carrying a good calf. I softly talked to them and mooed which intrigued them and caused the more courageous or inquisitive ones to come a step or two forward.

With all of them lined up now, counting was easy. I counted the front legs, divided by two, and got the right number. To check if I was right, I counted all legs, all udders, and tails and divided by six. I was still right. Having looked them over, I turned around and walked away, following the tabletop ridge which is part of the Flint Hills.

I stopped near a couple of old buffalo wallows. There are several along the hilltops but here there were about five close together, each about the size of a twelve by twelve foot room. Each time I pass them, I stop. But now I stood there longer, trying to see what it must have been like when the buffalo grazed these hills. Because we did not burn, the color of the range grass was still brown-green--the brown was the color of the dead grasses which had overwintered. The green was the new bluestem poking up throughout. But the buffalo wallow was the deep green of sedge grasses.

There were five of these deep green shallow, marsh-like, circular puddles--slightly dished out with standing water, no more than eight inches deep. Frogs were croaking all around me. This must have been a favorite spot of the buffalo. On top of the plateau with the breeze blowing to keep flies away. The old buffalo knew what they were doing all right.

Looking at the wallows I wondered how old they might be. Over time, they filled up even though they still show their shape and different vegetation grows in the compacted and wet site. At one time, I expect that the wallows were larger and somewhat deeper.

The buffalo used to wallow in these shallow pools and cover themselves with mud which provided a slick coating and must have given protection against insect bites. In summer when the shallow puddles were dry, they would have taken dust baths. I can just see it, the 1,000 to 1,500-pound bodies rolling and grunting; pure pleasure. As they rolled and covered themselves with mud, embedding dust seeds of grasses and other flora commonly found around the wallows, these then were carried away and deposited in the next wallow. There is a special flora which thrives around buffalo wallows in the compacted shallow dish of earth, dry in the summer and oozing soft mud during the wet seasons.

I regret that the domestic cows don’t treat the wallows with any respect. They’ll walk through it--not realizing they walk over the Plain’s history. They are ignorant animals held together by the barbed wire fence strung all over the hills.

From where I stood gazing at the shallow pools of water with the sedge grasses, I could not see a fence. All I saw was the far horizon, the low clouds, and I tried to see buffalos--hundreds, thousands of them. I know they were there at one time.

August 26, 1987

Cedar Tree

As you drive through our state, especially the eastern parts, quite often you can see a large tree growing near a gate or corner of a field. Those trees are landmarks.

It must have only taken a second for the sharp chain of the chainsaw to bite through the cedar tree's base. This spring, I bought a couple of new chains for the chainsaw to make cutting stout posts out of the old hedgerow easier. My sons used some of those posts a week ago when they restrung a boundary fence. Having worked together with them, they knew how I wanted it done. So, I left the job to them, thinking that it would be good for the two of them to work together. They worked well and got the job done. As a matter of fact, they may have done it too well.

Actually, what happened is they cut a tree I did not want to be cut down because I liked it and had plans for it. The funny thing is that there are plenty of cedars growing on the hills and one of the jobs is to cut all these cedars down. They compete for good grass. I figure where cedars grow, no grass grows and if you add up all the cedars scattered over a couple of hundred acres, you may be surprised how many acres of grazing are lost.

It is not only the economics. I don't like to see them scattered over the hills. They don't belong there. The cedars break the gently flowing lines of the clean sweep of the Flint Hills and that is how I want it to look--a clean sweep over the hills with only grass and cows. I'll let some brush grow up in the gullies. It provides cover for game and in a place like that, the deep green of a few cedars looks all right. It can still be debated from the management point of view as being an unwise decision because those few remaining trees can be seed trees. But that is the price I am willing to pay.

So, the boys knew that we were going to clean the hills and fence line of cedars. I can understand that once they got the chainsaw going and began to walk the fence line it became a thrill to cut down everything in sight, including that one cedar, my cedar, their cedar. The only trouble was, they did not realize how to look at and appreciate that particular cedar tree.

I had told my younger son earlier when we walked the fence line, "Hey, by the way, don't ever cut this cedar tree. I want it. I like it." I don't think I told him why and that might have been my mistake. Because if I had, he might have warned his older brother as they cleaned the fence line not to cut that tree. But he still might not have said anything because the tree was not standing in the fence line. It was about twenty feet away standing near the corner of the field, not crowding anything, yet it was a landmark. From way off, one could point to that tree and say, "There is the corner of our property and that is where the gate is."

And there was another reason why I wanted to keep it. On entering the pasture, it was a good tree where you could stand or sit in the shade or tie up a horse. Yes, the lower branches had been brushed up and needed to be cut off which I would not have minded doing. It showed gray cedar bark with knots. But the tree is gone. Like I said, the chainsaw must have roared through it like a hot knife slicing butter. He even cut it up in firewood length.

The son to whom I spoke about saving the tree said he never expected his brother to move that far from the fence line when cutting brush so hadn't told him what I said. "I'm sorry," he said. My older son felt bad for having done it because he realized too late that the old man liked the tree. Of course, you don't get mad when a thing like that happens, but I do wonder why it happens. It is not only a lack of communication, but a lack in seeing, understanding, appreciating, and imagining.

Maybe that is it. Not "seeing" the horse which is not there, standing under the tree on three legs, slowly swishing its tail, resting in the shade. Not "seeing" yourself sitting down for a moment and enjoying the place. Looking at the landscape, that tree had a reason to be there, even if it was a chance seedling allowed to grow up. That tree in the corner of the pasture, near the road, near the gate, provided a bold, deep green foreground to the gently rising hills in the background. It provided depth to the landscape perspective and you have to see it to be able to explain it.

Well, maybe I'll dig a hole and plant a tree there this fall! In the meantime, learn to "see" things. As you drive through our state, especially the eastern part, you can often see a large tree growing near a gate or corner of a field. Those trees are landmarks.

May 29, 2013

Modern Art

Every now and then, I walk into the Beach Museum on the K-State University campus and slowly move through the halls, stopping in front of a painting that either moves or intrigues me. I am not well versed in the arts, but I do know what I like. I also know what I do not grasp: some modern paintings where squares or circles are the main theme. I don't get it. Sometimes all you see are lines at different angles. It is art, but I do not stand very long in front of it before moving on.

The paintings on the walls in our home are realistic. They show what was there. I recognize the landscapes or the scenes. Several are from the old country, coming from our parents' homes and we treasure them. Many are already spoken for by the children. I remember when our youngest son, Martien, as a toddler stood in front of an old painting of woods behind a wide-open field of heather. A small trickle of water in a stream runs along the front. The sky is partly overcast. As he looked at it, he said, "I really like that. I really do." The light fall must have been just right, it spoke to him.

We have a large modern painting given to us as a housewarming gift by a friend who is an artist. The colors are deep and beautiful. She painted it at a difficult time in her life but that was all she would say. It was for us to visually explore the painting. I am sure each of us sees something different when we look at it. Often, the best way to see a picture is to take it off the wall and then re-hang it again. Too often, one gets used to what hangs on the wall.

Anneke, who is a docent at the Beach Museum, has stacks of art books that help her with the interpretation of art and to show young or older students what to look for. I have looked through several of the books when the snow was blowing and a good log burned in the fireplace. And just like in the museum, when the art was realistic, a landscape or seascape, I would stop to look at it. Those images made up of colorful squares and lines I tended to pass over quickly. Sometimes the colors would hold my attention briefly but understand them, no. My imagination could give it meaning but that would then be my interpretation. Maybe that was what the artist wanted or expected.

But the other evening that changed as I saw something unfold in front of my eyes that looked like a modern painting. As a matter of fact, afterward, I took a blank sheet of white paper and drew and colored in what I had seen. Then I saw I had a modern painting. I'll try to describe it to you as I'm sure you have seen it also. The difference may be that I witnessed it through a big sliding glass door, framing what I saw like a picture.

We have had some stormy weather and that late afternoon, dark clouds moved in from the west. They were dark, fat, rolling clouds--all black-purple. You could look under these dark clouds to the last of the day clouds in the distance. But what was striking was that the evening sun was shining from above and behind the dark clouds, momentarily outlining the rim of the fading day clouds. The result was a straight deep orange line from left to right. Very, very striking. The orange line was set off by the faded blue-gray sky. Upwards into the storm clouds and below the orange clouds was gray sky resting on the now dark Flint Hills. It was beautiful. The last rays reflected on the line of the graying clouds.

To top it all off, where all these lines were horizontal, a rain squall drew slightly arching lines from the black clouds down to the dark earth. It was all far enough away that it was a distant scene, just like a painting. I sat there in my chair and slowly watched this piece of art go by. I saw the rain squall move and as the cloud moved, the orange light faded. But there was a moment, the ephemeral, that I saw it all come together.

When it had all faded, I held it in my mind and went to my drawing table and took out a clean piece of paper. Quickly drawing a few thin lines where the different masses had been, I then got my colors. Black for the storm clouds. By pushing on the pencil, I made black, black clouds. I also grabbed dark purple. A mean-looking sky as I had just witnessed, and right below that, came a strip of blue-gray. Then there was the long row of clouds which had been highlighted by the setting sun. I had the right color, orange. And orange it was. Last, there were the long strips of gray resting on the dark Flint Hills. Wow! It surely was not art but it did look very much like what I had seen. With my black pen, I drew in the rain squall being pushed by the wind, hanging out of the black storm clouds reaching for the Flint Hills.

It was exactly like modern art except this was, or had been, real! And this was Kansas.

September 15, 1993

Alcove Spring

I thought I heard the jarring of wagon wheels and the complaining of the loaded wagon…

I have looked at the Niagara waterfalls, the Bridal Veil Falls, and other famous places where tourists gather to watch water tumble down steep, rocky inclines and wonder in awe about nature's forces. But I have been more in awe seeing the small trickle of rushing water coming from Alcove Spring some six and a half miles north of Blue Rapids, a community along the Oregon Trail.

Don't get me wrong, I think the Niagara waterfalls are awesome, but for me to have enjoyed them would have been before, well before, all the development took place. Today, safety barriers and chain-link fences are absolutely necessary for crowd control. The parking places, traffic control, and souvenir shops are all part of the ugly fanfare and I hate it.

My wife and I stopped at Niagara Falls six or seven years ago and we both still remember it as not a very happy moment. I did not want to stay around. For me, it was too commercialized. She, on the other hand, wanted to really look and see. So, we compromised.

I withdrew and watched the spectacle of people from a distance. From where I sat, I could hear the roar and watched the mist of the falling water. However, only when closing my eyes could I sense some of the wonder early explorers must have had when they stumbled upon this spectacular waterfall. It is gone for me. I have no desire to ride the boat. I have no desire to walk behind the falls in a yellow raincoat.

I tried to understand myself. I think it is the commercialization of nature in a very concentrated spot. It is like looking through a peephole and I don't want to do that. I am not against tourists--people who travel to learn about their country, their culture, and history. But, at Niagara Falls, I found it nearly profane. I will never go back.

However, I will go back again and again to Alcove Spring, which was a stop on the Oregon Trail. Edwin Bryant wrote in his journal on May 27, 1846, and I quote, "About three-fourths of a mile from our camp we found a large spring of water, as cold and pure as if it had just been melted from ice, altogether it is one of the most romantic spots I ever saw."

By the time people who followed the Oregon Trail reached the Big Blue River, they had been traveling about fourteen days, having started in Independence, Missouri. The 2,200-mile journey ended, if everything went as planned, when they reached the Willamette Valley in Oregon with its promises of free, fertile land.

If the Big Blue was swollen because of recent rains, wagons halted, and animals and people rested near Alcove Spring. The times I have stopped at the spring, there has not been anyone there, not that I expected anybody. I could walk the short distance from the road and follow a cow path and wagon trail among the woody shrubs winding down to the spring. All I could hear was the Blue Jay calling, warning other wildlife, "an intruder--an intruder--an intruder."

I carefully walked down so as not to damage any stone or plant. I sat down on a rock and watched the trickle of water, flowing as it has done for hundreds of years. The shade of the oak trees partly hid the cool spring. Very solemnly, I dipped my hands in the running water coming out from the side of the low bank. Behind the spring, off to the side, is the overhanging rock with dates and names carved by some of those who paused on their way west. The migration west started around 1841, gained momentum by 1843, and continued for the next 23 years or so.

Many travelers must have drunk and carried water from the spring, which supposedly has never gone dry. The Native American would have known about the ever-flowing spring and quenched his or her thirst from the flowing water. The wild game turkey, deer, and buffalo would have come to drink as well.

Over the years, trees have invaded along the stream down near the spring, changing the landscape from a more open prairie to a wooded, secluded spot. I sat on a stone and listened to the silence, the humming of insects, the bird calls, and every so often, the calling of a cow--buffalo. Did I hear a wagon with oxen come along the trail and the crack of the bullwhip to encourage the oxen to pull the last few hundred yards and then stop to make camp, rest, and drink clean water?

I thought I heard the jarring of wagon wheels and the complaining of the loaded wagon. Alcove Spring, a very quiet place.

September 24, 1997

Hay for Gus

After World War II, I heard stories from an Australian friend of mine who was stationed in England as a pilot of a Dornier--a plane which had floating capabilities, landing and taking off on water. This plane was used to look out for German U boats patrolling the North Sea. Anyway, he and his Australian buddies would get homesick for Australia and would visit the garden of a monastery close to where they were stationed and pick the leaves of a eucalyptus tree growing there.

Undoubtedly, the tree was brought to England as a novelty from the once faraway colony of Australia. They would crush the leaves and deeply inhale the aroma of the scented leaves and for a moment, they could think they were back home, walking in the bush. I know this aroma, the scent of the eucalyptus. It is very typical in the Australian bush.

However, for me it was a new scent that I inhaled deeply when I lived there. But it remained always an acquired scent. I had not grown up with it. I, on the other hand, could long for the scent of fall leaves piled deeply on the forest floor from maples, oaks, linden, and beech trees--the typical European hardwoods. And don't forget the pines.

Being young and busy in Australia, I did not have much time to be homesick. But I do remember, at times, recalling the European forests with their own signature scent. I remember very well when my wife and I sailed back to Europe that the boat we were on pulled into the harbor of Madeira, an island off the coast of Portugal. Standing on the deck, you could see deep green pine forests on the hills behind the city of Funchal. The warm winds sent a whiff of the pine scent to me. While standing on the deck, I took a deep breath, expanding my lungs, and sensed that I was getting closer to home.

It took another week of sailing before the steamer pulled into the harbor of Rotterdam and we were home. I walked the Dutch woodland and kicked the leaves and it all had a deeper meaning to me. I was back in the landscape I knew, which was part of me.

We returned to Australia and later came to the United States. From what I knew, I felt I would be at home here because the landscape, at least along the east coast, the Appalachian Mountains, was a landscape of deciduous and evergreen trees I would recognize. But above all, there would be the same familiar forest smell, especially in the fall when leaves drop and start to decay on the forest floor.

So, smell and scent are very strong elements and can help recall a familiar landscape. Having said that, just imagine my surprise when I was handed a somewhat bulky package by my wife's brother, who, together with his wife, is visiting us from the Netherlands. This is their first visit ever to us in more than 40 years. It is hard to believe. We have seen each other regularly and our families know each other well from our return visits to the Netherlands.

As is often the case when you visit someone, you bring something as a gift, a token of love or appreciation. And even though I had repeatedly told them just to come, they felt some surprises should come out of their suitcases. I must honestly admit that I am a difficult person to surprise. I always tell people, "Just come. Your coming is what makes the day." But they wanted to bring something that would make me really happy--but what?

So, here we sat around the table with cups of coffee poured and I was given the package. It felt soft and looked and felt like a big sweater--squeezable. I thought, "I already have a sweater. I like the old sweaters I have. I don't need another sweater and definitely not a new one."

As I handled the package, I noticed a familiar sweet, aromatic scent and said so. I watched their faces as they watched mine. As I unwrapped the soft package further, the aroma became stronger, more tempting, and more familiar. When I finally cut the corner of the large plastic bag, the sweet scent hit me like the salt spray of an ocean beach. It was the sweet aroma of the finest meadow hay grown around where I grew up. I pushed my nose into the bag filled with hay and I inhaled deeply.

There, in the bag, was the landscape of back home--the farms, the plowed fields, but above all, the meadows with the big oaks. The sheds with their heavy thatched roofs, the villages, the small steeples, the countryside where I grew up brought to me in the form of aromatic hay. What a great gift--what a great idea.

Earlier this summer, my brother-in-law and his wife were driving through the countryside when the idea struck her, "We are getting hay for Gus."

My brother-in-law, at first, did not think it was a good idea at all. But after a few miles, he too thought, "That is it."

They turned around, approached the farmer, and told the story of why they wanted only so much hay--only a bag full and not a whole bale. And when they told him who it was for and that the hay would go to America, the farmer said, "Oh, I remember his father well." And that completes the story of a landscape, of people--people who love the land, people who love each other--family!

January 3, 2001

Think Globally, but Act Locally

I wished I knew how old the grandfather clock really is. I know it is old. The clock just chimed six o'clock. It is the last six o'clock for the year 2000.

I am sitting in front of a roaring fire that has been burning the whole day. As a matter of fact, it has been burning for the last ten days. With plenty of wood, we enjoy the open fireplace even if it is not the most efficient way to heat a house. But it does pull people close. I noticed when the house was filled with family and friends over Christmas, that after people hung their coats, they gravitated to the burning fireplace with logs crackling as flames reached out. Outside, dusk quickly became night. The sun had already set when I went out to feed the horses. All there was left was a red streak poking into the gray sky, the last sunset for the year 2000.

Now sitting near the fire, I can reflect on the year that passed. As years come and go, it was not an easy year. A highlight was to make a trip to the Netherlands with our youngest daughter and her family to show her husband and the two grandsons where we grew up and had lived. It was a great trip.

So, what will the year 2001 bring? No one knows. The world is changing. People have said that before, but the saying "Think globally, but act locally" has an undertone for dramatic changes to come, which at Christmas time momentarily seem to come to a halt. I say momentarily because the traditional lighted tree, the burning candles, and the togetherness of three generations seemed to stop the clock for the moment--but not for long. And as you listened to the old clock, the tick-tock-tick-tock-tick-tock did not stop. The pendulum kept on swinging time away, minute by minute.

I am listening to it as I am sitting in front of the fire. It is very quiet in the house. A little while ago, I had a phone call from the Netherlands. The clock there had already struck twelve and so the New Year had begun. Remember, there is a seven-hour time difference because the sun rises in the east. They are seven hours ahead of us! Just a phone call, no longer expensive and you can talk. It beats e-mail. It is the voice and the background noises which make it instant. Think globally! I must call some friends in Australia tomorrow, but now remember that we are ahead of them. I should not wake them up after a night on the warm beach; it is summer down-under.

The world is getting smaller and therefore more interesting. Our youngest son did not come home for Christmas. He and a friend went to an out of the way place in Mexico. They were on the beach, had rented a jeep, and were twenty minutes away from a telephone. When he called home, he shared that they had climbed a Maya ruin and had quietly sat on the top looking out over the jungle. It had been very quiet, just what he needed. I am very thankful he sat there for a while. In my mind, I try to visualize the landscape--the ruins--the history of the place--the sky and again the landscape--the old culture. In his way, he was thinking globally, but acting locally.

As the years go by, the world will seem to get smaller and smaller. As we need to interact more and more, we become more and more dependent on each other, like it or not. Strange as it may sound to you, I saw this happen under the Christmas tree. As the family gathered to open Christmas presents, I took note of the origin of the gifts. I will list some and it should make you think. There were toys made in Taiwan, a jewelry box made in China, a belt made in Italy, a woolen sweater made in China but with an Icelandic design pattern, gloves made in China, Japanese film, even the wrapping paper was made overseas. Books were printed overseas. I was glad to notice that tools sometimes were still made in the U.S.A.!

But let's not look only under the Christmas tree. We have imported cattle from around the globe to breed with other breeds who themselves came from other places--Scotland, England, France, the Netherlands. We have looked for horses among the European bloodlines. While we are discussing animals, we might as well include people--they move, settle, and become part of us. I did! Don't you hear my accent?

All these things make it seem that the world gets smaller. Above all, it is people traveling, visiting places, and other cultures which seem to reduce the size of our world. However, where we can wrap our Christmas gift made in China in paper made in Taiwan, one thing we cannot wrap is--our landscape. Of course, we can see it in a movie with Fuji film or perhaps Kodak film, probably also made somewhere overseas. But the real landscape we cannot package. I know Walt Disney would love to try but it is impossible.

Therefore, for me the world is still very large with all its varied landscapes of deserts, forests, lakes, beaches, mountains, jungles, prairies, and plains. The natural landscape and the cultural landscape which, over time, people have had a hand in changing.

If you want to act locally, you need to try to treat the land with respect. It is all we've got--thank God we can't wrap it!

November 2, 1994

Empty Canvas

Do you remember having to write a paper for English and staring at a blank piece of paper--staring and staring--having no ideas and unable to focus? Or, as an artist, looking at an empty canvas and getting ready to draw the first lines, paint the first strokes?

I thought about that the other day when I met with a young couple and looked over their recently bought land, six and a half acres of empty wheatland. The lot was an odd corner a farmer had been willing to sell to straighten out his wheat field. But that is all it was: a wheat field with a slight slope to the north, otherwise flat with no trees, only a few power poles along the road.

Of course, the lot had air space. Geese flew over it. And in the soil, good soil, the normal soil activities took place. Worms crawled, beetles "beetled," mice "miced," and if they had been there, gophers "gophered." But above the ground, there was no activity. Only the strong south wind blew, which gave me a place to start--the first line to draw on the painting of this empty canvas.

"Where have you planned the front door of your future house?" I asked.

"On the front, to the south, I suppose," she said.

I immediately responded with, "All right, but how would you feel if this strong southerly wind blew into your home every time you opened the front door?" It was a thought she had not really considered.

"How far from the road do you want to build? What about the traffic noise and privacy? Where will the garage be--on the east side or west side, or offset? How can we make the best use of the lot to give you the privacy you want, but also make living here interesting and fun?" I questioned.

They wanted a small barn for some animals. He wanted a workshop. As we talked, inspiration started to come. The first lines on the canvas had been drawn. Oh yes, they still could be erased. We could still tear it all up, but we had started to fill the empty canvas, the six and a half acres.

The garage was moved from the east to the west because we didn't want to block the morning sun and moving the garage to the west made it much easier to use an existing long driveway. Keeping cold winter winds in mind, we moved the livestock barn to the west, opening it up to the south and east. The paddocks would run to the north half of the rectangular property. We drew on a piece of paper, rough--very rough, not even to scale--the concept of how this empty wheat field could be transformed into a home place.

It was obvious that the design for a shed and stables was easy--their function would decide their form. But the design of the house was another matter. Culture, tradition, holds its influence and also, of course, budget. Building a home is not cheap, never was.

But a home is our shell, and our shell can tell something about us. We have doors to step in and step out. What is it like to step out into the cold, snow, wet rain? What is it like to step into a warm, cozy, comfortable, spacious home with good furniture, a sturdy table to lean on, and share a cup of coffee and leftover pumpkin pie? The house has windows to look in, but also to look out and most importantly, to let light in--cheerful light, not the hot, blinding light of the 100 degrees Fahrenheit summer days. Closing curtains is not the answer--good design is. Especially when looking at where to position the house properly on the land.

As we stood in the middle of the empty field, there were only a few landmarks in the surrounding wide-open space. Facing south, way in the distance, stood a beautiful old barn and further west, a farm with buildings and large trees. Both elements in the landscape could perhaps be borrowed as attractive views, distant views. With any future tree and shrub plantings, these two views should be considered.

Of course, the sunrise I already mentioned but I don't want to forget the dark nights, the full moon, and starry nights--all part of our landscape. It is sad that barnyard lights, artificial light, handy as it is, takes away the sharp detail of Kansas starry nights. I have a switch on my yard light so when I want to, I can switch it off and enjoy the silent moon and starry night without interference.

To help with planning, you should draw your lot to scale and also cut out to scale the home, barns, and other structures to be built. Place these cutouts on the lot and move them so that roads, parking, traffic pattern to the barn all make sense. Just the other day, I saw a new home on a large, attractive lot but the parking space was designated right in front of the house--awful. It ruined the look of this otherwise attractive home.

Having placed all structures to their best advantage--keeping climate, view, drifting snow, traffic patterns in mind--you can start to plan on paper for tree and shrub plantings, leaving space for the yard, a lawn, play area for small children, and a vegetable garden. As the trees mature, they will transform the once empty wheat field into an attractive home site with character, just like we look to the farmstead with a windbreak and mature shade trees. Living and growing on a home place you planned provides you with a very special feeling and meaning to life.

Someday, you'll drive home, enter the driveway and you'll recognize it all. You will sense that you have turned an empty wheat field into a place to call home. You have turned space into a sense of place. You "painted" your home on the canvas of the Kansas landscape.

December 27, 2000

Trails in the Snow

They flew in from Raleigh, North Carolina, landed at Kansas City airport, and rented a van to drive home to Manhattan. Between the two parents, grandparents, and a sister with her family, the grownups and kids slept, ate, and played. It had been a typical family Christmas, whatever is typical with extended family trying to do it all, perhaps trying to do too much. It all worked out though.

Yesterday evening, a row of wet socks was hanging in front of the burning fireplace. They got wet when the kids went sledding down the hill because we did have snow. Remember when two weeks ago I said, "I hope it will snow when you go out and cut your Christmas tree?" Well, when I went out to cut ours it was snowing! And ever since, the landscape has been white with a little more snow, added every now and then, making it a real white Christmas.

The main roads and interstate were clear, but county gravel roads had a cover of snow which made driving them a quiet pleasure. I drove the three-mile county road to the farm slowly, enjoying the winter scenery of the familiar countryside. However, familiar as it may be, the snowfall made it look different. Some of the hills looked closer, while others along the horizon seemed to fade into the gray sky.

Seeing Black Angus cattle grazing, a cluster of black cows standing around the freshly rolled out hay bales made a clear black-on-white picture. On the other hand, our two white horses tended to blend in with the snow, only their faces with their soft, dark eyes stood out. The same with trees and fence posts, the rough bark on the dark side of the posts stood out against the clean white snow.

Walking the prairie with my son, we did not talk much but enjoyed the silence and the winter scenery from the vast expanse of the snow-covered Flint Hills to the small, singular standing grasses with their delicate seed pods and the coarse seed pods on the sumacs. There were dried seedpods from flowers. Some bent over and you could see how the wind had blown the seedpod to and fro as if one were sweeping. The result being a very fine etching, a few lines in the snow.

As we studied the fresh and old animal tracks, we both became aware of our own big feet tramping through the snow. There were tracks going everywhere--deer, pheasant, and quail. The quail tracks were interesting because you could see how, as a small covey, they had scurried from shelter to shelter. They seemed to come together near a tree trunk and then scurry forward like infantry advancing in enemy territory. The quail should not feel that pressured, but they do. As we walked on, we disturbed a covey of them along the wooded edge of a field. The quail exploded into the air and settled down a hundred yards ahead of us.

Seeing all the different animal tracks, I once more became aware of how many different animals, large and small, live on our farm in its diverse landscape. It is the trails that tell us about their lives, or the end of it. The other day I read an interesting observation by Donald Stokes in his book *Guide to Nature in Winter,* and I quote, "More than just an impression in the snow, a trail is like a rope connecting disparate moments of time--both a record of the past and a connection to the present, for back on the trail one sketched the encounters of a living being, while ahead is the present animal continually leaving its life experience one step behind."

The point became very clear to us when we walked the farm road. There we noticed the tracks of a vole, probably a packrat, crossing the path. It came out of some low, dark shrubs on the one side going to the now snow-covered hay meadow on the other side. However, the tiny paw prints stopped in the middle of the road. It was clearly visible in the snow what happened.

Where the paw prints stopped, there were two imprints in the snow just like children make when they form snow angels. They fall on their backs and move their arms along their bodies, creating angel wings. What we saw imprinted in the snow were two wing imprints, real wings, most likely from a red-tailed hawk. A hawk which had circled overhead, saw the rodent, swooped down, and caught it. The animal never made it to the other side of the road. It was lifted up and undoubtedly eaten by the hawk. Such is life!

November 28, 2001

Great Plains Prairie Postage Stamp

It is only a 34-cent stamp, but even that one small postage stamp portrays the complexity of the prairie. The whole sheet of stamps depicting the Great Plains Prairie is like a puzzle. Several times after having bought the whole sheet, I wanted to send friends the remaining border after having lifted and used the individual stamps. Even the edge shows our prairie in all of its beauty, diversity, and energy.

There are three parts to the sheet: the sky, the prairie surface, and the underground part that is close to the surface. The sky is alive with a threatening thunderstorm. Dark clouds with lightning streaks push thick, fat, rolling white clouds ahead of it into a still, quiet blue sky. There is a lot of wind and rain in that storm. It does not take much to "hear" the booming thunder way in the distance. You know you better get home quick because you are going to get wet.

Sun rays are still greening up the prairie, but as the storm pushes forward, all will turn dark. A Swainson's hawk swoops and is buffeted by the first surges of wind while a small covey of quail moves with fast wing beats to find shelter. Canada geese, flying into the storm, seem to be veering like an airplane avoiding the turbulence of a storm. There is a lot of activity in the sky.

Buffalo are grazing on the prairie, as yet they are facing the storm. However, I expect them soon to turn around and put their short tails into the wind and hang their massive heads low, weathering the storm. The pronghorn in the foreground are looking out across the wide plains as if trying to decide what to do. The distant lightning seems to have caught their attention as if they know that lightning can set prairie afire. Yet being the fastest running animal, they can outrun a fire. This scene, this prairie landscape, is of a scale that there is space to run.

Very strange that it is all captured on a few postage stamps which together create the vast space--the puzzle! The strong horizontal lines of the prairie and the skyline are broken by a few vertical lines in the foreground. There is the upright purple prairie clover with its interesting flower: the cone and fuzzy purple base. Prairie coneflower and little bluestem send their roots deep down into the prairie in the foreground. You can see the roots, deep roots, go down--down. Soil activity in the prairie is teeming with bacteria breaking matter down, assisted by beetles, mice, gophers, rats, and ants. With all the animal activity, the dung beetles are very, very busy.

In the chain of life, I look at the prairie rattlesnake hiding in a burrow underground. I look at the hawk. Both will snatch a mouse or rat feasting on grass seeds and other vegetation. I am just waiting for the meadowlark to explode out of the taller grasses, to fly up--up--up, and proclaim it is spring or a beautiful summer day. Why a summer day? Well, because the grasses are deep green and the flowers are blooming. On the other hand, it may be late summer as the grasses show a seed head.

There is a lot more to be noticed and learned from this composition of postage stamps, making up the third in a series called Nature of America. The next time you need postage stamps, buy this whole sheet. Having bought it, look at it and flip it over to learn what you are looking at. To me, the whole sheet of stamps is a composite of the prairie. It is a rich, very rich, habitat.

Three short paragraphs provide a brief overview of our prairie landscape, and I quote, "The varied tapestry of the tall-grass, mixed-grass, and short-grass prairies reaches from the eastern woodlands and oak savannas to the foothills of the Rockies. Grasses and wildflowers make good use of limited rainfall, and fire helps sustain the ecosystem. Prairies provide habitats for many animals, including the pronghorn--North America's fastest land animal--and the prairie dog, one of many burrowing animals that live on the prairies.

Explorers were impressed by the immensity of the central and western grasslands, the region that came to be known as the Great Plains. Settlers' steel plows altered the landscape and transformed life on the prairies.

Native prairie is rare today; remaining patches exist because of careful management and diligent preservation efforts. But the defining characteristics of the prairies live on; for instance, cattle, rather than bison, are now the dominant grazing animals of the plains."

Our landscape--you can see it on a postage stamp.

Vernaculars

Tell me stories. Where have you been?

July 31, 1985

Old Boots I

There is a country road just north of Russell, Kansas, where I must go to bury my cowboy boots.

I don't know about you, but all clothes are not the same to me. I know there are people who collect caps. You can walk into their office or basement and the shelves are lined with caps--John Deere, Co-op, Milwaukee, seed companies, etc. On the collector's head, an old, greasy cap holds his hair down or hides the fact that he has little hair left.

I don't collect clothes as a collection to show off. My problem is I collect them because I don't throw the old ones away. Some of my clothes, and I am talking about field clothes, are truly worn out. My wife would have dumped an old Levi jacket years ago, but she knows better. She has even bought me a new one, but the old one is my favorite with worn buttonholes, torn collar, oil stains that no longer wash out. Yes, jackets take a special place in my wardrobe and the older the better.

I am not a shirt fancier. I like them to fit, but otherwise, I am easy and hold few sentimental attachments to shirts. The reason may be that old shirts never die. They become rags you grab to polish the car or change the oil. I have recognized rags as once having been on my back, a shirt worn on a camping trip, or on a trip overseas. So, I get weaned off my old shirts while working with them and with little regret do I throw them away when a rag is too oily to clean. Jeans also wear out slowly. You can cut the legs off and make good sailing shorts and that way they literally fade away.

But boots are a different story. Boots are different from shoes. I can throw old shoes away without any remorse. I like to wear shoes out but once they are too old and worn, they go out to the dump. I can't do that with boots. So, what do you do with boots? Boots so worn that the soles have broken and have let loose because the stitching is gone and there is no leather to hold new stitching. I have boots that are a sandy gray because they have not seen any polish or oil for years. There are now cracks on the calf of the boots. Still, I keep them and I can't throw them away.

They have been on my feet while I visited the old country, walking in Norway and in Scotland. I remember when I bought them. I kept them polished and they stayed clean, worn under suits or jeans. They have gone to weddings and they have gone to funerals. I have had hot feet in them and cold feet. I just cannot throw them away. The leather straps have been cut off--one was already broken--and I have saved the brass rings, which is the typical dress up for the Shenandoah boot.

Hearing this, the real boot fancier may laugh at me and say, "Boy, just throw those things away. They are not even in the class." And he may be right with his pointed pigskin, crocodile, kangaroo, or rattlesnake boots. But, so what? I am attached to mine and hold my memories dear and I just can't toss them in the trash can. But I also know that I can't keep them standing on the shelf forever. Sooner or later, they'll be hidden by cobwebs.

However, I think I have found the solution and it is a simple one. What strikes me is that other souls must have had the same problem I have but saw fit to solve the problem in a way that openly showed their sentiment. They too just could not throw worn out boots aside, but wanted to give them a place of honor, to be looked at by the passerby--a boot cemetery or eternal resting place with soles turned up.

The next time I am close to Russell, Kansas, I am taking my boots along and I'll drive just north of Russell and put my boots there with all the other upside-down boots on the old fence posts near the gully north of town.

August 7, 1985

Old Boots II

Last week I shared with you my problem of discarding old cowboy boots and how I'm going to solve the problem by doing what other people have done before me, sticking the boots upside down on fence posts. There is already a row of boots, a long row, along a dirt road north of Russell, Kansas. I passed that road once on an evening at dusk and pulled off the road to take a closer look at the upside-down, worn out, dried out, cracked boots. All were different styles, but they were all cowboy boots.

This fall, I am going to stick my boots at the end of that row and there they will remain. The fall rains will soak them and the wind will dry them out. A spider may crawl in and find protection. The winter snows will cover the cracked soles. Horses walking the fence may stop for a moment and sniff the leather, snort, and go on. Cattle may walk up and rasp the old leather with their rough tongues, but the boots will stay upside down on the gnarled fence posts just as all the others have.

You wonder if boots could talk, what stories they would tell each other. Looking at all the boots, I think a lot of stories could be heard. Some stories about bad crops, the day they walked into the field after the hailstorm and kicked the truck tires in disgust. Some probably remember how they stepped into the bank to arrange a loan or to pay off a debt. Some boots probably remember how they went out courting in the new pickup truck and how they walked quietly into the house when they came home too late. Some, undoubtedly, have rodeo tales of kicking bulls and racing horses. Some probably talk of the day the horse put its big foot on them and what was said.

All would have tales of long days in dusty or muddy fields, working the land and growing crops, of tending cattle. They all remember heat and cold and tired feet. Of men or women straining to pull them off and setting them, the boots, down with a clunk on the porch or kitchen floor.

The boots all remember how at first, they were treated as someone special, but as they got older, they were used more, worn every day, and how that was fine with them. As a matter of fact, they probably all liked it because now they were real boots and could prove how tough they were.

I hope my pair of boots won't feel too bad when they have to listen to all those stories. Mine can't brag about rodeos, driving big tractors, or working with cattle. But my boots can talk about other things. They have walked the streets of Paris and the narrow lanes of old towns in countries far away. They have walked the roads where I grew up. They have seen Dutch cows, windmills, and tulips. They have heard people say things they did not understand but they liked it anyway.

On second thought, I sure hope my old boots are not going to talk too much. I hope they know when to shut up and listen because where they will be resting on the fence post, there is a lot to hear besides boots talking. There are the crickets, the quail, and pheasants calling. There is the wind blowing around the bluff and blowing through the grass and weeds along the fence line. The coyote walking by and the cattle grazing in the distance or coming close to lie in the cool shade close to where my old boots will be resting on a fence post near the draw along a dirt road north of Russell, Kansas.

June 26, 2002

The Old Wire Gate

You can see Yankee gates all over the Flint Hills, some good and some not so good. They hold cattle as well as any bought metal gate, if not better. It has been two years since the five children and their spouses got together and decided to give me gates--proper gates--to enter the farm. They had put a collage together of photos and pictures of what it was they wanted to give, and it did look functional. It was two swinging gates attached to sturdy poles. Indeed, it would make driving into the farm easier as a swinging gate would swing aside when entering.

However, I was attached to the old, rusted, barbed wire gate! The wire gate had served many farmers and as people had tripped over it or had broken the thin posts, it had been repaired, wires spliced, and yes, even some rusted baling wire had been used. We have all seen those wire gates, leaning and holding cattle.

It is true, everyone complained about the entrance gate to the farm. But I ignored it and since I am the one who opens the gate most of the time, I just left it as it was. Each time that I opened it, I pulled it, swung it around, and leaned into the bushes well out of the way so that anyone following me would not drive over the poles.

Wire gates are cheap to make yet, if you work at it, they can be a piece of functional art. Another name for them is Yankee gates. Of course, you should know that the word Yankee has a Dutch connection. The early Dutch settlers had names like Jan and Kees. Both names were, and still are, popular names for men in the Netherlands. When English-speaking people listened to the early Dutch settlers, they constantly heard the names Jan and Kees and before long, they coined a new word, "Jankees," which became Yankee.

The word Yankee is associated with people living in the northeast who are thrifty. I must admit that when the kids wanted to give me the more functional gate for the farm, I thought that the old wire gate was good enough and would last for several more years and if needed, could be easily fixed.

Well, even good Yankee gates wear out and my old farm gate did too. Someone ran over the pole and someone else got hooked on a wire. For a while, the gate was in shambles. The bottom wire was dragging and the looped wire which held the gate upright slipped off and then the whole gate would collapse. It looked cheap.

It was high time to do something. I was not going for a fancy swinging gate. Yes, I know the kids wanted to give me that and I don't want to sound ungrateful, but I decided another Yankee gate would be just fine. Only, I would build a really good one. I quickly located three five-foot-long poles from Osage orange trees that were nice and straight. I cut them and tossed them in the truck with tools, barbed wire, as well as the chainsaw, and Kit, my grandson, and we were off. The dogs came along for the ride and to see if they perhaps could catch a packrat or chase a deer.

We stopped at the old gate, pulled through, and quickly tore the old wire gate down. While Kit rolled up the old gate poles and rusted wire, I started the new construction. It is easily done and goes quickly. And what I like most is it doesn't cost much. We all have pieces or a roll of barbed wire. If you have woods or an old hedgerow, you can find straight poles if you look. The only thing is that I wanted heavier poles. But that is all right. It makes for a stronger gate.

We only used three wires. With a pole in the middle, that is plenty. I even strung a vertical cross-wire between each pole. To finish the project, I made a good latch so that it is easy to take the pressure off the gate and open it. When we were finished, Kit and I stepped back and briefly admired our handy work. It will be there for a long time and I hope he too will open the gate many times. Later, we built another one between the hay meadow and the horse pasture. It too is a piece of art with three sturdy poles holding up the wire frame.

Now Kit knows how to build a wire gate. The only thing he said was he could not build it as quickly. That does not matter, Kit. I like it that you now know how to build a good Yankee gate. A gate which will last for many years and every time you open it, you will know you built it. And when you don't open it, but quickly try to crawl through it and tear your shirt--you know what to blame--the cheap Yankee gate!

July 16, 1997

Harness

I know what I am telling you has happened to many farmers, however, to hear Bud tell the story made it special. It all happened a long time ago, but the twinkle in Bud's eye showed me that he remembered the incident as if it happened yesterday.

The incident came back to Bud because we were standing on the spot where it all happened--that was before the creek had cut out soil eight feet deep and twelve feet wide, causing a dry creek. Of course, when it rains, this now dry creek runs with water coming from the grassy hills. The time Bud was talking about, the creek was just a sway, a low spot in the field through which you could plow, harrow, and pull the lister, planting corn. All this was done by a team of horses. Bud remembered those days. He enjoyed working with horses.

The simple fact that at one time our land was worked by horses is a given, but proof of that was the aged, dry harness hanging in the old barn which burned down a few years ago. Along the wall, you could see the straps and other hardware which, at one time, belonged to the harness for a team. Earlier I had saved some of the parts as mementos.

I also have worked with horses. I remember the sound of working leather and the smell of horse sweat, horses standing, resting at the end of the furrow. Or at lunch break, with their heads down--just resting, slowly eating oats out of a bag.

It all came back to me when a few days ago I walked the dry creek bed. Looking down as I slowly walked, I picked up a stone and let it drop again. I found an old metal casting but it was buried too deep; pulling on it did not move it. The gravel held it in its grip. It always amazes me what you can find among the rock and stone, and there is always a story behind it.

As I slowly walked on, suddenly I saw a metal form which looked familiar. The lime of the limestone had coated the piece so that it blended in perfectly, except for its form. I stopped and picked it up for closer examination. For its small size, it was heavy. Most of it was gone--rotted, rusted. But this part was still whole. It was the top part, the ornamental part of a harness; the rigid piece for the heavy horse collar to which the traces were attached for pulling. This one was solid brass and it was heavy. I have some hanging in my workshop, but those are the simple ones without ornamentation. This piece was a classic--the solid brass told me that it has cost good money.

As I stood in the bottom of the creek gently rubbing the brass, I wondered what horses had been wearing it--they had plowed the same fields I now owned. They had pulled wagons to go to town, probably Westmoreland. Oh, I wished that piece of the old harness could talk and tell me about what was and had been.

Much I can interpret just like an archeologist at an archeological dig. Because I generally know, but there is so much I do not know. Like Bud told me how, at one time, the creek followed a different course and the fields we now walk and work had a different configuration. If the brass piece which I held in my hand could talk, it could have told me.

But, if the old piece of harness couldn't talk, Bud would share. As we were standing at the edge of the field, he told me of how, years ago, the John Deere tractor he had bought had been delivered to the same field. At that time, he had been working with a team of horses using a lister, planting corn. When the tractor had been brought to the field, they unhitched the team and put the new tractor in front of the lister. He was a young man then. Even today he is still young in spirit. You only have to look into his eyes and hear his laughter. But it is true, he is getting on in age--past 80 now.

Anyway, when he started the tractor--a John Deere with a hand clutch--he came to the edge of the field and as he was turning, he hooked the lister into some wild grapes which were growing rampant. As the long canes of the grapes grabbed and entangled themselves around the machine, things got out of hand. The only response Bud had was, "whoa, whoa, whoa," but the new John Deere just kept going, pulling the vines along. Only when Bud pulled the hand clutch did the new contraption stop.

Oh, I wished I had been there. But to hear Bud tell the story, you were there. You could see Bud's consternation that the machine did not stop on command like the horses did. Oh Bud, it is a great story. Every time I think of it I see it happen, and whenever I work or pass the same spot, I will remember--"whoa, whoa, whoa,"--and will hear you laugh and see the twinkle in your eyes.

August 2, 2006

The Last Apple

I walked several times around the apple tree and looked at the damage the fire blight, a disease, had done to the structure of the tree. Large branches had turned black and bark was peeling while leaves had dried up and crumbled. The crotch of the tree, where the grandchildren felt safe to climb and sit, was infected. The virus disease had struck all three major branches. Yet, the tree was sprouting new branches, deep green branches, from the lower part on up. There were even a few apples hanging on.

However, I knew the tree was doomed and decided to cut the apple tree down while my grandson could help me haul off the branches to the growing brush pile. I did not like to do it as the tree had a prominent spot in our yard and had been a tree to play in for the children. So much so that I did not realize it until my granddaughter, Joni, stepped into the house and looked out of the window and noticed the tree gone. There was an empty spot, a vacuum, some roots, threads to a childhood suddenly gone with the roar of the chainsaw, severed forever.

Later, when we shared a meal together, Joni and I joked about it. Actually, it was not joking. It was expressing deep-felt sentiment about a gorgeous apple tree which in the spring was loaded with blossoms. But, more so, had been a place for her to climb away from everybody and find a place to sit in the safe crotch of three sturdy branches.

I knew that as the tree grew, I would prune the branches so that they would provide the needed shade for the large sandbox. When quite young, her older brother would play in the sandbox while the pony would stand next to him eating apples. When the children played together, they would climb the tree and vie for the perfect sitting place, the crotch. At times they would call each other names--"my place--no, my place--no!"

When Joni's mother was a teenager and we lived in Blacksburg, Virginia, she too had a climbing tree with her favorite reading spot. That was an oak tree, a large oak standing along the side of Toms Creek which ran beside the old farmhouse where we lived. A place we loved.

Having made my decision that the dying apple tree should be cut down, Matthew and I set out for the task. With one pull, the chainsaw roared and while at first I very selectively cut the dead and dying branches, I quickly came to the realization there wouldn't be anything left of this tree. As I cut the tree down, I did cut firewood out of the longer branches and the stump while Matthew hauled the brush off. I finished, leaving a twenty-inch stump which I cut off perfectly level. On top of it, I placed a large water dish with a good-sized stone in it. It made the perfect birdbath, something my wife always wanted. I never could think of a natural one. However, this was it.

The last thing we did was rake all the twigs and small branches together and threw them on top of the brush pile. When I looked at the job, it was well done. Well done until Joni came in. "Baba, you cut my childhood tree down, my security, Baba." Instantly, I felt as if someone whipped me. "How could you, Baba?"

I realized that deep down in the near subconscious, this apple tree near the sandbox held meaning for Joni. Somewhat bantering, yet serious, I apologized to Joni and said that it had hurt me just as much to do it, that I had agonized over the decision. I had spoken with the tree spirit and it had conceded.

We bantered for quite a while. Then, suddenly, she said, "Baba, I need some paper and a pencil." She found some on my desk, walked out, and sat down near the stump. She was composing a song--a song. I watched her through the window, leaving her creative spirit alone. But for a moment, I watched her beat out a rhythm with one hand while sitting there.

Later, we walked together to the stump and noticed some shoots which I had left standing. These were promising shoots coming up from the strong roots. Those shoots coming from the roots will not be like the original apple tree, as it was grafted on seedling rootstock. The new shoots are unknown promises. They are wildlings full of vigor. They will grow into a tree, a tree large enough to climb or to sit in its shade.

While cleaning up around the stump, we must have missed the small apple lying to the side. Insects, bees, or a bird had started to peck into the apple, but more than half of a solid apple was left. I picked it up and put it in my pocket. Back inside, I washed it and said to Joni that she and I should share that last apple--and so we did. She cut it in half and I peeled it. Then we solemnly chewed our halves. The apple was not yet ripe, but having some blush it did have a slight flavor. It was a good apple, the last apple. A tiny seed saved from that apple lies on my desk.

This fall we will have some applewood to throw on the irons in the fireplace. By then, all the brush will have gone up in smoke and we'll have danced around the burning pile like a Maypole to celebrate the life of this apple tree.

August 9, 1995

The Stone

Our route took us through Iowa, then north through Minnesota into Canada by Thunder Bay. From there we drove north again to the "lonely road" which runs along the southern edge of the northern frontier of Ontario, Canada. It was a beautiful road with forests and lakes on both sides.

Once we entered Quebec, we drove further east on Route 117, then higher north on Highway 40. With a wide arc, we at last drove to the city of Quebec and south again to Montreal. From Montreal, we drove south to Saratoga Springs, New York, where my daughter, husband, and two grandsons now live.

They are building a new home on the most interesting ten acres I have ever seen. It is a diverse, beautiful site with all the trimmings. A small gurgling stream just long enough to call some of the clean water your own, large maple trees, and many white pines all separated by small open fields, meadows. All of this held together by the remnants of old, stone walls. Stone walls made of round fieldstones unearthed when farmers cleared the small meadows for cultivation. No wonder the fields are small. One would only want to carry those big boulders a short distance. Farming must have been very hard for those early settlers.

On a short walk my daughter and I made down to the small meadow, we startled 22 turkeys which upon our entering the open field, marched off cackling like a clan of kindergarten kids on their first outing. They disappeared under the pines along the edge of the field with one taking wing and flying along the edge of the woods just a hundred yards from where the new home is being built. But I am running ahead of my story.

A few days before we left, I told my friend and colleague, with whom I would be traveling, I had a favor to ask but to please be honest and say "no" if he didn't want to. "Okay," he said. "I will--what is the favor?"

I felt somewhat sheepish but continued and said, "I want to take a big rock to my daughter. It's a rock from Kansas, actually a rock from our farm."

"How big is the rock?" he asked. I told him that I could just lift it. He said, "Okay, that is fine. We'll put it in the car when I pick you up."

The evening before we left, I drove to the farm and down to the creek where sometime earlier I had found a beautiful eroded, large stone. With much effort, I had lifted it from the creek bed and up the bank to leave it there for later use. This was the best use I could think of--to give it to my daughter and son-in-law for their new home as a small touch from Kansas. It is a beautiful stone sculpted by water erosion over time. I saw all of the Flint Hills in this one stone--the ridges and the valleys.

After I backed the pickup close to where the stone was lying, I walked over and hoisted the stone into the pickup truck. As I was huffing and puffing, I wondered what my friend would say once he saw the rock. After all, he was driving only a small Toyota station wagon. But the deal was--he would tell me honestly.

I drove home and quickly packed clothes so I would be ready the next morning. When Steve drove up, I looked into his already packed car and was about to forget about my stone when he said, "Where is the rock?"

I hesitantly opened the tailgate of my truck and watched Steve appraise the large stone, but he did not flinch. I kept my hopes up and asked him where he thought it would go. With a little adjustment of suitcases and other stuff, we made room on the rear seat. I placed some large, empty feed bags over the seat so as not to wear the fabric and proceeded to lift the big stone out of the pickup and stagger to the car. I carefully let the rock down on the rear seat where it stayed, never moved, the next eight days while we were on the road.

As we drove the northern route, I saw many rocks along the road and among the trees. As we drove into New York from the north, there were boulders--big, round boulders--all deposited during the ice age. I saw the boulders as part of the landscape among all the rocky outcrops so typical of the Adirondacks. But as we got closer to the new home for my rock, something dawned on me as we drove up the newly cleared, old farm lane to the site where they are building. I saw all the large boulders and the rock wall on their property and I suddenly realized, "That is all they need--one more rock from Papa."

We stopped at the end of the long, wooded driveway near their temporary quarters, a large camper, and I unloaded my gear. After I had done that, I said, "I brought you all something from the farm," and somewhat sheepishly I hoisted my rock from the seat and carefully laid it in the grass next to a huge boulder my son-in-law had unearthed when they placed the camper. My heavy rock looked very small, but it was a Kansas rock--from our farm, from the creek--and it was a beautifully shaped stone. My daughter said, "Papa, that rock comes in the house--all the other stones stay outside. This stone will go with us wherever we go--this is a Kansas stone."

September 29, 1999

The Little Red Barn

Driving slowly by, I remind myself to bring my camera next time and take a picture of the little barn. For more than a decade I have driven by it, but just the other day the setting of this little farm building really struck me. It was the golden color of the ripening soybeans in the foreground which set it off.

The small barn is getting old. Having only looked at it from the outside, I expect it is more of a storage building than a barn. But the red paint makes it a barn for me. The small square windows still show some white paint where the sash hangs in the frame. The whole building could use some fresh paint, but I doubt it will ever get it. It is an old barn.

Around it, tall weeds and grasses have grown up and soften the corners. Wild purple asters and goldenrods grow mixed in with some tall grasses. Behind the barn is a stand of smooth sumac showing off its gorgeous, deep red, fall color which reminds me of a red wine. Actually, the color of the leaves blend in with the old red of the barn--autumn is here.

What made it all stand out was the golden field of soybeans changing color day by day as the beans ripen. Right now, the leaves are still showing that rich, golden yellow color so typical before they turn brown, dry up, and start falling off. I know if I want to take that picture I should not wait too long. Golden, yellow-colored beans, red sumac, purple rough-leaved dogwood, and the old red barn. Next to the barn, a small tree is growing. Someday the tree will push the barn over and then the old building will be a memory.

The weather-beaten barn reminds me of an old, familiar face. As it ages, the face reflects peace with the world. It no longer shows the assertive expressions, the urgency, and the quick temper flashes. Behind the barn stood a mule in a small fenced-off paddock. The hedge posts held up the wire. Further behind the paddock is an old shelterbelt and wood lot growing toward the top of the hill. The clear, blue sky held it all together.

The last time I looked at the barn, I kept my eyes open for the flock of turkeys that roam in those parts. A few days earlier, I surprised them while they were sunning on the south side of a hill. I was walking down the hill following the fence line. Had I been hunting, I would have made a clean shot. But I got more pleasure from seeing them be surprised and scurry away. A few took to the air and flew off for a short distance toward the red barn. Others scuttled away in the tall grasses. Standing high, I could follow them easily with the sun shining on their dark feathers. They clucked as they moved away from me. Together with the old, red barn, they made the picture complete.

Robin Langley Sommer put a beautiful book together titled, *The Old Barn Book*. Flipping the pages, I see big barns, small barns, red barns, white barns, stone barns, and weather-beaten barns. They tell a story just like the old face I talked about.

If you know farming, you can easily fill in the stories the barns tell. All you have to do is quietly listen and observe. The few fence posts still standing tell you what animals moved around or were housed in the barn--chickens, hogs, cattle, horses. Each had its own fencing needs. If you stepped inside the barn, you might still see some old harnesses hanging on a peg. The leather all dried-up and cracked. But at one time, the barn was used to stable horses or simply to store tack. Many smaller barns were used for storage or for chickens.

I don't know the story of this little barn, there must have been a house close by and people walked from the house to the barn to gather eggs. They stood inside when it rained or snowed, and it kept them dry and out of the wind. They painted it. The good soil beside it must have been cropland for many years. And it just happened to be that this year, beans were growing in the field and I happened to see it when the sun rays hit it just right--the perfect picture.

November 9, 2005

The Old Board

Out in the shed, I keep neatly stacked lumber; being of the generation which either wants to save or recycle material that may have a second, if not third or fourth, use. Yes, I also have three buckets of old, rusty nails, having dug often in them to find the nails I needed to mend or fix something. However, I don't do that as often anymore. And when I looked at the buckets full of nails, I thought about their next best use--digging a two-foot-deep hole on three sides of a tree showing a bad case of iron deficiency and dumping a bucket of these old, rusty nails in each hole.

I have found that the old nails are of better quality than the new ones you buy in fancy, expensive boxes with a see-through, plastic window. The old nails are stronger. Just yesterday, pulling nails out of an old two-by-twelve-inch board, I caught myself thinking, for a moment, to save the straight nails. But then on second thought, I tossed them in the wide-open dog food bag in which I deliberately toss what I do not want or need to keep; making an effort to dispose of the nails so they don't lie around on the garage floor and cause a flat tire.

But the two-by-twelve-inch board of redwood I saved is something else, with no knots and the growth rings very close together. A beautiful piece of wood you would not find today in a lumber yard. Just looking and handling that piece of wood started me thinking, "Where had the tree grown? How tall a tree and how old before it was cut?"

The piece of wood I held was from at least 50 years ago. We had used it as the big step up and out of the garage to our patio. I wondered how many footsteps this wooden step had carried. The patio is now gone and instead, we have a very useful deck.

I wanted to build a big workbench in the garage so that woodworking would be easier and I could get things organized. For me, the thrill is to build something well but without needing to go to the lumber yard. I had plenty of wood to choose from, sorting out the needed wood from under the shed--the four-by-fours for legs, the two-by-sixes, and two-by-fours. It now became a matter of deciding carefully which piece of wood to choose for what.

The one heavy board had a story to tell. It was redwood without knots or blemishes. However, it did have nail holes and holes drilled for bolts to hold hinges. That heavy board really spoke to me. Again, I do not know where it came from. A few years ago, my son helped to take an old barn down and I ended up storing some of the big lumber. But I didn't think this board was part of the old barn. Yet, it might have been lying around the old barn and been the side board of a horse-drawn wagon--heavy but strong. The reason I thought that is because it had been painted a rusty red. There were also old holes visible outlining where hinges fitted.

I put that board aside as being ideal to run along the full width of the front of the workbench I intended to build. Such a heavy, strong board, well-supported, makes working at the bench a pleasure. It feels good.

The other boards also have histories. There is one crusted by cement and concrete that must have been used in construction at one time. After running the old sander over it, the wood turned smooth but was still a rough piece of lumber. Excellent for the top of the workbench but not the first board.

Studying the growth rings of the front board, they are very, very close together. Quite different when you compare that with lumber grown on today's plantation. Nor is there the hidden history. I know how plantation lumber is grown and treated. There is not much to tell. The tree is planted as a seedling, grown as a crop on a plantation as fast as it is feasible. Then it is harvested, cut, and sawn into utility lumber to be used for construction.

But the piece I held in my hand and used for my workbench grew slowly, somewhere along the Pacific. The tree had grown in the mist and silence of the old forest. Maybe it should never have been cut; left standing.

As I walked by the workbench where it holds the honored place of the front board, I looked at it and saw many past uses. I heard the horses pull the wagon through the cornfield and the corn cobs hit the catch board as the wagon moved forward. As the men quickly pulled the cobs and tossed them in, the men told stories and joked. It was a good crop and the wagon soon filled, pulled by the horses out onto the road to be unloaded in the corn crib next to the old barn--happening a long time ago.

I wished that the board could talk!

September 20, 1995

Love Knots

People sometimes think I make stories up, but I don't have to!

It's the golden time of the year in the Flint Hills. The goldenrods are blooming now and the cottonwood leaves and grasses have turned to shades of amber. It is the time of the year for gold, but the gold colors nearly fooled me the other day.

Last Thursday evening, my wife suddenly realized she had lost one of her golden earrings. She had happened upon a single earring mysteriously lying on the floor and after much searching, she found the gold back of the other but couldn't find the earring itself. I had given the earrings, a pair of "love knots," to her years ago and she wears them regularly, showing that she still loves me. But now, the house was in a panic--where was the golden earring? They are not that big and it could have been swept up quite easily, rattling up in the hose of the vacuum cleaner, and then it would have been gone--gone. My wife felt quite miserable. After all, they were--love knots.

With only two of us at home, there was no need to start blaming anyone. I don't wear earrings, nor does our dog, a nine-month German shepherd, Kansa. Of course, playful as a young dog is, he likes to chew on anything--sticks, toys, etc. He concentrates on sticks which he proudly carries into the house, and of course, his knotted rope. The rope he treats like a dead animal, flinging it around in all his fury.

If all this sounds rather wild to you, this is the household where, when the parents were gone, the children brought the pony into the house through the open doors of the living room, into the dining room, and made it feel at home--feeding him grain at the table while everyone else enjoyed donuts and coffee. They recorded this event with a photo which they sent to us while we were on vacation with the comment, "Everything is fine. We are doing great. Don't worry." So, a dog is just another critter in a home that loves animals.

So, where was the golden earring? My wife kept looking--looking everywhere. I never realized that there was gold, real gold to be found in the Flint Hills. Yes, there are the goldenrods and the golden leaves, and their color nearly fooled me. I heard it against my shovel but thought that the yellow was a small piece of golden-colored leaf and that what I struck was no more than a stone, a piece of flint. Now I know how a lucky miner feels when he finally thinks he's struck gold. When I turned on the garden hose and I carefully started to slush the shovel, like panning for gold, what I thought was perhaps a leaf, got form and was yellow. It was the metal--gold all right.

Carefully, I kept washing so the spray would not blast the small nugget away to be lost forever. The more I washed away, the surer I was that the small nugget was the lost earring and I knew why we could not find it. We had been petting the thief. Kansa runs outside in the fields but just this once he relieved himself on the lawn and luckily, I decided to clean it up. As I picked it up with the shovel, my gold-panning pan, instead of tossing it with a big swing over the fence into the pasture, I had decided, at the spur of the moment, to take a closer look.

When the small earring lay clean and shining on the shovel, I picked it up carefully and examined it. The dog had crushed it slightly but didn't destroy it. He must have swallowed it like he swallows pieces of sticks.

After further thorough cleaning, the earring is back in the jewelry box. As a matter of fact, it has been worn again. After all, they are--love knots. The moral of the story is: there is gold where you least expect it. And if I see you walking the Flint Hills with your nose to the ground, you know what? I won't blame you because I found gold--real gold where I least expected it, and it was not fool's gold--it was a lost golden love knot.

A version of this story was previously published in a church newsletter.

November 17, 2004

The Old Scoop

When I found the old scoop, it was lying among dusty, broken boards and other useless junk left over from an earlier time. When the barn was still used as a barn, when tools would have been set up against the wall or hung on a nail or peg. I had walked past the scoop, stepping carefully, watching so I would not puncture my feet on a rusty nail protruding from a fallen board. I thought it could be dangerous to walk in the old barn with its collapsed loft and badly leaning posts with old, dusty straw bales lying around. Yet it all intrigued me.

I was not expecting to find anything of value, least of all the faded memories of people who had farmed the land and used the barn. There were the horse stables, actually bays, separated with a swinging beam, a pole. Except they were no longer swinging. With the upper floor sagging, the beams were now laying on the dirt floor.

Opposite the stables had been the corn bins and that is where I noticed the old scoop. It was rusted from many years of no use. But even though the scoop was rusted, it still felt solid and when I knocked it against an old foundation stone, the scoop had a ring to it as it was made of metal, a real scoop, not one of those plastic modern ones. It felt solid enough that I took it with me into the daylight for a closer look.

It was a good scoop, rusted but with no cracks or dents. No one had ever run over it. On one corner, the metal was slightly bent upwards as if it had repeatedly been struck against a protruding nail or board while unloading the wagon. I decided to keep it, taking it home to find a proper handle as it was the handle that was missing. Of course, on an old farm tool like a scoop, the handle can tell stories as well as the scoop.

Like I said, one corner of the old scoop was slightly bent up as if it repeatedly had struck something. I wonder who had used the scoop and very importantly, who had used it last and left it lying in the corner. Who broke the handle? Who had leaned on it looking out over the fields?

To use a scoop, you bend over and let the scoop do the work. As it moves over the smooth wooden floor or concrete to pick up, better said to scoop up, to load, or unload the grain. Who had leaned on the scoop, straightening his tired back? As I held the scoop, I wondered how many bushels of wheat, oats, and corn my scoop would have thrown on the wagon or off the wagon into the bin.

The more I thought about the scoop, its passing life, the more I liked it and I was happy to have salvaged it. Sometime later, I was in Hays in Ellis County and while there, stopped in at the old hardware store--a real hardware store. There I found a scoop handle with the right curve and a firm "D" grip which I bought. I believe it was the last one. The owner was not going to stock anymore. There was no demand for scoop handles. He was closing the store--going out of business. We talked and reminisced for a while and I went on my way, holding the handle tightly. Closing the door behind me, I heard the old bell ringing me out. It was the last time I stepped into the old store--a real hardware store.

The new handle hung for several years on a peg on my tool rack while the scoop remained leaning in the corner of my garage. There was no hurry to put the two together. Besides, I would have to treat the new wooden handle to fit the curved neck of the scoop. I knew how to do that but never felt the urge until a few months ago.

The trick to bending the wooden handle is to heat it in boiling water. One morning, I put a large pan filled with water on the stove. Once the water boiled, I placed the last twelve or so inches of the wooden handle in the pan and let it sit for a good five or eight minutes. When I thought it had boiled enough, I took it out and immediately rammed it into the curved neck of the scoop. I kept driving it in until it could go no further.

It was a near-perfect fit with just enough tilt to the "D" handle so that working the scoop would be a pleasure. The metal quickly took up the heat of the wood as I practiced the smooth movement of scooping air. It felt good and I knew I had found myself a good scoop.

All I had left to do was drill two holes through the neck and secure the handle with two cut-off nails, flattening the cut-off ends, and beating it with my hammer so no sharp edges remained. I made sure that when the scoop ran through my hands, I would not cut myself.

All done, I hung the scoop on the peg beside my other tools. Actually, I hung it behind my short-handled shovel, also a good tool. It is like a scoop, but smaller. And while we are at it, let me note that a shovel is not a spade. The scoop is in a class all by itself. A scoop is used for shoveling grain. Its sides curve upward to hold a load and the size of the blade is large.

By now, I have used my scoop and I was happy to feel how well it worked. I was not standing on a wagon pulled by a team of horses unloading oats. No, I used the scoop to load wood chips which had been dumped at the end of my driveway where I park.

For two days I loaded my wheelbarrow with chips and pushed it to the border where I dumped the load to spread with a fork or my hands between the shrubs and perennials. I have forgotten the number of wheelbarrows I pushed, but I know it took only about five or six good scoop loads to fill the wheelbarrow. I smoothly pushed the scoop under the wood chips--one, two, three, four, five--six.

After two days of moving wood chips, I was glad to hang my old scoop back on its peg. But it felt good to use it and I enjoyed the work. I could see the slow progress made, scoop after scoop and wheelbarrow after wheelbarrow. But I must admit to being very glad that I did not have to unload a big semi-trailer loaded with soybeans with my scoop--now that would have been a real nightmare!

December 6, 1995

The Double Trunked Tree

"Grandma asked me to remind you to come out and look at the tree--the two trunked tree. You and she had talked about it years ago when you gave a program at Randolph." Indeed, that was years ago and it once more reminded me how fast time flies and how busy one is.

And now many years later, I was talking to Kiffnie, the granddaughter, who is a student at Kansas State University. The first time I met her grandmother, Kiffnie must have been a youngster riding a pony on the family farm. She knows about the tree close to the swimming hole in the running creek.

Months went by and a week ago she stopped in my office and said again, "We should go and look at Grandma's tree." Right then and there, I grabbed my calendar and circled the date, an open spot on my busy calendar.

When we finally drove out this past Friday afternoon, the weather was beautiful. Actually, the place was not far--a 30-minute drive just north of town. The last miles are gravel road and then you turn off into a wide valley with a stream running through it and meadows along the sides. With a good snowfall, it would make a perfect Christmas postcard. I instantly fell in love with the place. There was so much landscape character.

A valley always holds a special charm. There is the element of surprise--down the crest, around the bend, across the bridge. There is nearly always rich soil and ample water. Remember, water carved the valleys here over millions of years. Some parts are often wooded with tall, magnificent trees.

The valley landscape is the classic of "void"--the wide-open space--surrounded by "mass,"--the hills, in this case, Kansas hills. A valley is a secure place to go. The family who now farms the valley was displaced when their old farm was flooded by Tuttle Lake. It must have been very difficult to leave earlier roots behind of land settled, developed, and cultivated by their ancestors in the big valley of the Little Blue River. But this small valley was theirs and the white clapboard fence around the farmstead spoke of happiness and pride in taking care of the land.

When I stopped the car and got out, I met both grandparents. Grandma reminded me that we had met a long time ago. I felt guilty for not taking the time to come out. And I was mad at myself for having missed the opportunity to see this part of Pottawatomie County earlier. I drive through it to go to other places but lately have come to realize that "Pott" County has many hidden spots as beautiful as our own farm.

After we briefly talked, we jumped in the pickup truck and drove out to the tree. We followed the valley where the horses grazed, galloping across the brome pasture when they heard the honking horn of the truck. They stood snorting near the electric fence, expecting some reward for dashing to us. All we could offer was some petting and love.

The dogs caught up with us and we drove on, crossed the flowing creek at the gravel bed, and turned into the woods to stop near the double trunked tree. It is an intriguing tree--one large tree with a smaller, but still good-sized, tree standing next to it. At a height of about eight to ten feet, the smaller tree bent over and literally continued to grow through the larger tree. It was as if, at one time, a hole was drilled through the larger tree and the small tree, then a sapling, was laid through the hole and everything grew together at the joint. It has resulted in an arch through which, at one time, a rider and a horse could ride but now only gave room for a man on foot.

Grandma took a photo of the tree, Kiffnie, and me. I treasure the photo. It reminds me of a moment, a moment a long time ago. A moment when something was done by whom I don't know, for what reason I don't know. Maybe it was done out of playfulness by a traveler who had wandered off the Oregon Trail.

I don't think it was an act of nature. It just cannot be explained that way. There still is the question of why so high on the trunk, well out of reach of one standing on the ground. Another question I could raise is did it point to a certain direction? Did someone bend the tree to hang a cooking pot over a fire?

When we walked back to the truck, we saw a flock of turkeys through a clearing in the woods. One of the dogs saw them too and ran after them. It was kind of funny when the turkeys took to their wings--scolding the dogs as they scattered in several directions. Maybe one of their ancestors was cooked in a pot hanging from the woven branch--slowly cooking over the fire. The photo on my desk reminds me of a riddle and my thoughts keep turning back to that spot in the valley with the two trunked tree.

September 16, 1998

Tie Rings

When you walk across the Kansas State University campus, there are many paths to follow. Over the years, more and more concrete has been poured. The sidewalks have become wider and in some places, thickened so that in an emergency or for efficiency, trucks can drive over the concrete where students normally walk to and from class.

Walking across campus, I often follow the sidewalks, weaving my way through the throngs of students who move from one building to another. However, sometimes I go off the beaten path. It takes me in-between the buildings, especially the older buildings which interest me. And as I take shortcuts, I see the buildings from the back. Over time, much has changed. Newer buildings have been added and the function of older buildings has changed.

However, if you take a good look, you can still recognize earlier uses of the older buildings. The doors and windows tell stories. You can recognize the old livestock buildings on campus where animals were housed. I expect mainly cattle. The stonework of some of the older buildings is more randomized, more typical of a barn structure. You can clearly see where, over time, different courses of stone were added and changes were made.

Passing behind Burt Hall, I always glance at the older stonework and enjoy its more vernacular architecture. You have to look for it and look beyond the outside metal staircases and other ugly additions that have been joined to the back of the buildings. It is like walking an alley.

The other day, once more passing by, I took a closer look at some old ironwork that protruded from the stone wall. It is nothing special, nevertheless interesting, and added to the stories these old buildings could tell. What I was looking at were old tie rings embedded in the wall. They, at one time, had been used to tie cattle for classes or simply grooming.

With the State Fair going on right now, my imagination visualized some Jersey cows tied to these rings. Their soft brown eyes looking at the wall while they stood there patiently, their good points being explained to the students. Next, the cows would be groomed and fussed over as only dairy cows could be. I say "could be" because with today's larger herds, cows don't receive the individual attention they did when herds were still small.

How do I know these rings were used for cows? Oh, I expect there might have been a horse tied to them at times. However, they are too low to tie a horse to, especially a young horse still impatient, turning, and stamping around. I would tie a horse higher to make sure it could not put its leg over the tie rope. These rings are about two and a half feet up the wall. Of course, over time, some soil may have been added. A little bit of digging could tell us that.

On a shelf in my workshop, I have a couple of similar tie rings attached to metal anchors which were embedded, at one time, in an old barn at the home place in the Netherlands. When that building was remodeled, the old tie rings were taken out. They were saved and I brought them with me. I hope that someday I can use them, put them on a structure, and once more tie livestock to them.

I wonder how many students, faculty for that matter, who walk by the Burt Hall have ever noticed the tie rings? I think if I were a student today, during some dark night I would find some good old cows and just for the sake of good old times, tie them to the ring, give them some hay, and watch what happens when people come to work and see a few cows in the parking lot!

January 17, 2001

They Don't Hang Right

It is especially at this time of the year that celebrated traditions become happy memories. It was more than ten years ago that I shared with you the following vignette on Christmas. I have often thought about the old lady because she was very special, as was her husband whom I highly respected.

I wonder if her grandchild, who was given the old sleigh bells, treasures them as much as the grandparents did. I think so and in doing so, I am sure that her grandparents live on as very happy memories.

"He gives his harness bells a shake to ask if there is some mistake."

"You can do something for me," she said, "open the door. You see the bells, they don't hang right."

When I approached the house, I had seen the harness strap with sleigh bells hanging over the front door. And as the door opened and I was invited in, I heard the soft tinkling of the bells. But now the lady of the house told me, "It's not hanging right." She added, "There should be a hole in the middle of the strap, and then it should hang evenly with two equal loops."

I took down the long leather strap with the bells, heavy bells jingling in the crisp air. Looking it over, I quickly saw that this was authentic, a real antique. The dark, leather strap was still supple, but you could see that it had been worn for many years, many snowfalls, many sleigh rides by the horses which pulled the sleigh. The bells were real sleigh bells with a joyous winter sound.

She told me, "We bought them in New York. We really could not afford it, but we liked them so much. It cost fifteen dollars and we bought them."

As she told me this, her small figure slightly bent over and her face, many, many winters old, shone with happy memories. Her whole face spoke of earlier Christmases. The first Christmas, these sleigh bells had decorated the front door of their home and invited friends and family to enter.

Now nearing the end of her life, her husband recently had died, she again decorated her home with the traditional and heirloom decorations. But the bells had to hang just right, and could I please do that for her. She watched me closely, giving directions on how it had always been done and that is how I did it, making sure that opening and closing the front door would not cause the sleigh bells to fall down.

I stepped back to see if the two sides hung evenly and counted the bells on each loop just to make sure. She asked, "Are they now ringing if you open the door?" So, I opened and closed the door, and the sleigh bells were ringing, and her face was happy.

Once more I stepped into the house, closing the door so the heat would not get out. I told her how, years ago, I had taken sleigh rides. "Me too," she said, "through the really deep snow."

I could see in her eyes as she gazed in the distance that she clearly remembered. She once more saw and felt the snow. She heard the jingling of the bells and heard the soft sound of horses' feet--plop, plop, plop, plop--as they pulled the sleigh through the snow. I think that sleigh ride had been 85 or 87 years ago.

If only I had remembered the poem by Robert Frost, "Stopping by Woods on a Snowy Evening." I would have softly recited it and I know she would have joined me, or perhaps only listened. But whatever, she would have felt as if she were sitting in the sleigh.

"Whose woods these are I think I know.
His house is in the village though;
He will not see me stopping here
To watch his woods fill up with snow.

My little horse must think it queer
To stop without a farmhouse near
Between the woods and frozen lake
The darkest evening of the year.

He gives his harness bells a shake
To ask if there is some mistake.
The only other sound's the sweep
Of easy wind and downy flake.

The woods are lovely, dark and deep,
But I have promises to keep,
And miles to go before I sleep,
And miles to go before I sleep."

Plop, plop, plop, plop, plop, plop, plop.

Reflections

The end of a Kansas day.

January 31, 2003

The Mouse and the Stone

How many Native American artifacts, I wonder, are plowed up on our former prairies to be buried again as we work our fields, never to be found.

His desktop was a large, smooth door supported by two low filing cabinets. The fact that you couldn't see the top of his desk was because of the two computers, the writing pads, and stacks and stacks of papers and books spread out on top of the desk, as well as on the floor of his study. When I entered, following him, I stopped at the door not knowing where to step. But he forged ahead, entered the room knowing where to step so as not to mix files or kick things out of order--his order! Seeing him walk to his desk, I carefully followed so I could see the computer screen on which he was trying to display some information he wanted me to read.

With his strong hand on the plastic mouse, he clicked and clicked again, trying to coerce the needed information from out of somewhere--out of the innards of the computer. A modern tool with more capabilities and memory than I will ever understand. I chuckled at the two computers on his large desktop, to which he explained, "That is my old one, but it holds so much information which I have not yet transferred to the other one; therefore, I cannot do without it." I expect that in the not-too-distant future, he may have three computers on his desk as more and more information is gathered and more articles are written to be stored, edited, and sent.

The computers as typewriters are the modern fountain pens--they take up desk space, but we can no longer do without them. And possibly, all the scattered papers, the information they contain, could be stored on a disk. His room would suddenly be an organized space, but oh, so sterile. One disk instead of all those flowing papers and books lying around, each breathing life--stories, tales, and facts.

It is when he moved his hand on the mouse that I noticed--the realization was sudden. The pressing papers must, at first, have hidden the simple yet beautiful object. Next to the mouse, I suddenly saw a stone, but not an ordinary stone. It was a stone somewhat shaped like a computer mouse, only slightly larger. It was a dark, very smooth stone with deep grooves along the top. It was an authentic ax or tomahawk. A tool, a weapon, at one time used by a Native American, probably some 3,000 years ago. The grooved stone looked like dark soapstone with deep veins leading inside. The very striking part was how much its shape resembled that of the mouse--the very modern next to the old. I want to say the very old but 3,000 years to an anthropologist is not that old.

His room holds many artifacts and old stones, tomahawks, and other prehistoric weapons or utensils. However, it was the smooth, dark ax or tomahawk stone lying next to the computer mouse which struck me as timeless.

Someday the mouse will be thrown out. Sooner or later, it will malfunction and no longer will be able to convey the commands given to it by the power of the click induced by the pressure of the finger. But the stone was so much like the mouse that without looking, you could have grasped it and given it the command to pull up information, to delete, or send.

That stone will last forever. It already has been a stone a long time, no ordinary stone. Whoever picked it up from the many other stones realized that he could, with hard work and other primitive tools, transform this smooth stone into a tool and/or weapon. The stone became the head for a tool or a tomahawk.

Who was the Native American who held it in his hand and after long, long polishing, found it good? He then attached it to a handle and he used it when, where, for what? If the time of its creation is estimated as archaic, what did our land look like two, three, seven thousand years ago?

At some time, the stone was lost--let go. Maybe he threw it, missed, and let it lie. Hard to believe, after all the polishing and hard work of making the groove. Over time, the wooden handle and leather strap which held the two together rotted away and the stone was left. Left for ages. It was picked up, looked at, then taken home, and treasured.

The stone is a link to the past, a past which an anthropologist who studies people and earlier time periods may understand. For me, the dark stone was a moment of wonder--a moment made very striking as the stone was lying next to the mouse of a modern-day computer.

Several years ago, when visiting an old farm in Norway, the farmer gently opened an antique cabinet and showed me a display of stone tools and weapons found on his farm while they worked the fields. Those tools were from the earlier feared Norsemen, the Vikings. I held some in my hands before placing them back in the cabinet.

Who are we? What will we leave behind--a plastic mouse?

January 28, 1998

See if You Can Grow a Tree

When the phone rang Sunday evening and I picked up the receiver, I learned that a colleague had died. We all expected it, only a miracle could have saved him and though you hope, at a certain moment you know that even a miracle won't do--that the end of this life is near. So, when the call came, I had expected it.

People die all the time--old, very old people, young, and the very young. It is when someone you know dies that the death has a greater impact. I have seen my share of death and I know that people, family, and friends go on living. They rebound and work through their grief and life becomes sweeter, at least for a short period.

I noticed it as I drove to work the day after the phone call. My thoughts were with John who had died, who was no longer suffering from the debilitating cancerous tumor in his brain. I looked at the morning sun, the clear sky, the birds, and all the other living things going on living. I thought about John's wife and children who now have to cope with death--which, in some ways, can be seen as a relief. The suffering and pain are over for John.

John leaves his mark. He was a horticulturist, a professional, and his impact on horticulture in Kansas has been well documented. However, it is not always the big things for which a man or person is remembered.

After I received the phone call that John had died, I stepped outside for a while and walked around the yard, taking note of all the things I still could see and the feelings I could associate with them. I looked at the horses and saw them watch me, ears perked up as if they were waiting for my whistle telling them to come. I just stood there watching them, rejoicing in the fact that I could. I was thankful. I looked at other things--the birds on the feeder. They were getting their last tidbits as the evening was falling.

Slowly walking on, I stopped by my favorite tree--a tree I planted more than twenty years ago. It was a gift from John, who now is dead. It actually had not been a tree yet. It was an acorn he and I saw lying at the edge of the compost pile at the research station in Wichita where he served as director. John said, as he carefully picked it up and held it in his hands like one would hold a young bird which had fallen out of the nest, "See if you can grow a tree out of it."

The acorn was a Bur oak. It had split and two tiny leaves and a sturdy white root about three inches long proved to us that this acorn wanted to live--wanted to become a tree. I carefully took it out of John's hands and placed it in some damp peat moss so I could take it home.

Once home, I laid it near a downspout to keep it moist, to be planted later. The next few days I was on the road. When I got back, I took a look at the sprouting acorn. It was still there. That weekend I planted this tiny miracle.

In the first years, as the sprouting acorn became a small tree, a few mishaps happened. One of the children mowed it off; the rabbits chewed it as it resprouted. Admiring the persistence of the tiny tree, I placed wire around the seedling and drove a small stake into the ground close to it so no one could say, "I did not see the tree, Papa." With the protection, the Bur oak took off. It is said that Bur oaks are slow growers. But with a small amount of fertilizer and water when needed, there was no stopping this tree.

One year, with the care I gave it, it grew four feet--the next year, three feet. I now had a small tree eight feet tall, and that was many years ago. Over the years, the tree kept growing, it lost its spindly form and put out stout branches. Each fall there is an increasing crop of acorns which the squirrels enjoy. Last fall, I picked some of the acorns to try to grow some offspring from this tree, John's tree.

Walking my yard last evening, I stopped at the foot of this Bur oak and gently touched the tree. I let my hand go over the rough bark. I stepped back and visually measured its height by increments of six feet, my height, and I reckon the tree is now 30 or more feet tall. It has a beautiful shape with a nice central leader.

Some years ago, my grandson, Kit, climbed to the top to remove a broken branch. He found it a little scary as the tree was swaying when he reached the pinnacle. However, he managed well and once back on the ground, I told him that I planted that tree as an acorn. He looked up and said, "Wow."

The young sturdy Bur oak holds many memories, but the first one is that John picked it up, held it in his hand, carefully handed it over to me, and said, "See if you can grow a tree out of it." That is a great memory.

January 19, 2011

The Old Clock Talks

Every winter I want to experience at least one period of deep snow. A celebration of winter with big snowflakes drifting down so that one can see the accumulation of snow falling on outdoor structures, giving them a different contour. There should be no wind or blizzard conditions. No, it should be just falling snow, thick and silent. And very importantly, it should snow the whole day but especially still at dusk. The evening should be white, a snowy evening, with silent snowflakes falling down.

Yes, I know I am selfish, utterly selfish, as I don't need to go anywhere nor do I need to drive home from somewhere in the snow. It is true I can enjoy the exceptional beauty of silent falling snow.

I do need to go outside to feed the horse but that is part of the beauty. Even though they have shelter, they would rather wait outside and whinny when they see me coming with their grain. Then they get their hay, some extra so they stay warm. Placing their hay inside the shelter, for a moment the snowfall is outside but as soon as I step back through the doorway, the silent snow is all around me again. The branches of the trees are loaded with snow, tops of wooden fence posts and rails are covered with snow. It is a wonderland and it is silent. I love it.

If you have cattle to feed, I hope you have the hay bales handy and that your tractors run and gates open easily. The good thing is that snow cover provides protection and moisture to the wheat crop. So maybe my selfishness is not too bad.

There is a reason why I enjoy such a silent snowy day. Why is it special to me? On such a silent evening when outside all sounds are muffled, inside the house is quiet and only the warm fire glows. It is then that you can hear the old clock. On quiet evenings my mother would say, "The old clock talks."

I think it happens when it is totally quiet that you suddenly sense this talking. It happened to me that snowy evening a few weeks ago. I had been reading and put the book aside, pulling up to the fire, and looking into the warm flames when all of a sudden, in the silence, I heard the old grandfather clock. I heard him talk. The talking comes from the special sounds the old and very simple timing mechanism makes, keeping time and chiming on the half and full hours.

The clock has been ticking for many, many years and in many different places. Except during the war. Then the heavy weights were taken down and carefully wrapped and buried in the garden. The Germans wanted the lead. The old clock said, "No way."

After the war, the weights were dug up and polished until the copper shone again. I have tried to visualize the moment when my mother touched the pendulum and once more the old clock talked, keeping time.

The clock traveled to Australia with my parents and later returned to the Netherlands. Moving a few more short distances before it came to hang in The Polhoeve, the family home place.

It was our oldest son who once said when the kids were discussing the future of the old clock, "I have seen and heard it run at Opa and Oma's at The Polhoeve whenever I visited as a child."

Upon which our youngest son said, "But I have heard it talk my whole life while I grew up here in Manhattan, in this house on the hill." Because after my parents died, the Grandfather clock was once more taken down, carefully packed, and sent overseas all the way to Kansas. It now hangs in our living room, keeping time and telling us the hours.

Our son-in-law, Wayne, loves clocks and keeps this one clean and running. The old mechanism intrigues him. I am sure that if someday a piece is worn out or breaks, he will be able to repair it. In my mind, I see him carefully cut new pieces for the simple mechanism out of a plate of metal. He has a patience and love for clocks.

As I was watching the flames in the fireplace, feeling its warmth, my mind kept drifting off, following the old clock in its moments of happiness and moments of sadness. I saw my mother standing there, pulling the heavy weight up-up-up, guiding the chain. Life goes on.

Across the top of the clock lies a hand-embroidered dust cloth. Tradition says every new owner should make a new one or else the clock won't keep time. But we have kept the one made by my mother and the clock doesn't seem to mind. Under the cloth which lies across the top of the clock, I placed an envelope with a photograph of Anneke, my wife, holding Martien, our youngest, as a baby in her arms, listening and watching while my mother pulls up the weight. A very happy moment in the home place, The Polhoeve. I know because the old clock told me so while I was sitting in front of the fire and outside the silent snow was drifting down.

February 7, 2001

You Cannot Rule After Your Death

A friend of mine has a saying, "You cannot rule after your death." This means that you can make plans which you in your own life may live by, but one should not make plans and rules which box in your children or grandchildren.

In principle, I agree with that idea. However, there are circumstances or situations where ruling after your death can be very interesting. Anneke, my wife, told me of such a situation and I will share the story she told me.

There was a man who lived in Hamburg, Germany, a good hundred and twenty years ago. When Kansas was being opened up, the man read about the opportunity of cheap land in Kansas and having always wanted to own land, made up his mind to go there. He packed, said his goodbyes, and started to travel. He sailed across the Atlantic, landed in New York, went undoubtedly through Ellis Island, and once outside the gate, looked west. I don't know how he traveled to Kansas, probably by train. Because it was the railroad companies that advertised in Europe about the Kansas land sales.

He came to Junction City and there he bought land. The land had to be settled. He had to build a house, a shed, and start fencing but he was now a rich man. He walked on his own land.

In the morning, he got up early and watched the sun rise over the hills. He listened to the meadowlark. I wonder if, by chance, he still saw a lone buffalo, a remnant of the vast herds. He must have watched the coyote and probably cussed it when he missed a chicken or two. He plowed his land and grew crops and the railway hauled it to the markets for him. Slowly, the land around him developed into more farms. People who had followed a similar dream were his neighbors.

To see what it must have been like, I look at old photographs and read the stories about other settlers, immigrants, who made this country--the Germans, the Swedes, the Norwegians, Finns, Italians, the Dutch, and many, many more. Marrying and raising families and tending the land, growing and losing crops. They learned that farming in Kansas was not easy, but they could own the land if they worked hard. Over time, they built better homes, better and bigger barns, as well as schools and churches.

When the old German died, he wrote a will. It is that will which has intrigued me and partly contradicts the saying "You cannot rule after your death." I cannot tell you much more about this German who left Hamburg to follow a dream. I can only tell you that by pure coincidence, the fourth-generation descendant of this German visited and stayed in my home place in the Netherlands as a friend of my daughter when they traveled in Europe together. The fourth generation is now in their early forties.

The fourth generation is significant in the long-dead German farmer's family because when old, the farmer ruled in his will that the farm could not be sold until the fourth generation! And the fourth generation is living now, raising their families, but not on the farm. They are professionals. By this, I don't mean to say that farmers are not professionals--not at all! But the fourth generation is not living on the farm. I don't even know if the six hundred acres or so are leased or farmed at all. I do know they have never been sold because of the ruling in the will.

Now, what does this mean and what does it tell us? A few things. Great, great grandfather was a wise man, a man who lived a full life. He was a thinking man. He loved Kansas. He loved his farm. He must have been a successful farmer because he did not have to sell his farm to provide for his old age. Again, there must be many things the family has recorded about this man of which I have no clue and are not important. Important to me is that the great, great grandfather experienced a sense of place, which he only was willing to let go after four generations.

He could not predict the future. The rise and fall and rise in land prices. He never thought beyond the family farm. Family farms could grow bigger if there were strong children to help on the farm.

I am sure the old German farmer believed in education because later generations pursued learning and professional careers. Yet today, the family still owns the land, and it is this generation of the 40-year-olds who now have the freedom to sell--to sell the farm. Great, great grandfather no longer rules.

Can this generation look four generations back? Can it look ahead? What will Kansas be like a good hundred years from now? What will people's dreams be? What will land, space, mean to people? What will it mean to people to know where they came from?

It is true that one should not rule after one's death, but you can surely make your family sit down four generations hence and ask them to think!

March 23, 2005

The Wooden Chest

When we have had visitors over from the Netherlands, we often have driven out to Goessel and visited the museum there. The town lies near Alexanderwohl, north of Newton. Goessel, according to what I read, was named for a sea captain whose ship had transported many of the Mennonites across the Atlantic. The ship sank at sea on its return voyage. Apparently, the captain was able to rescue some of the passengers, but he went down with the ship.

Inside the museum, I like to spend time viewing the mural which depicts the arrival of a group of Mennonites on the prairie and the process of unloading the train. Of course, there is no station. The train just stopped on the tracks in the middle of the prairie and families gathered around their belongings set down among the now trampled prairie grasses. It is the faces of the people which interest me. The determined men, seen handling the large wooden shipping boxes which hold the family belongings.

Whenever you look at an old painting or a mural of people moving, especially the immigrants, you see sturdy wooden boxes like very large suitcases standing on the wharf or, as in the case of the immigrants settling Goessel, on the prairie. If you visit the Kauffman Museum in Newton, you can see wooden boxes depicted in the display.

The same can be seen in St. Louis if you visit the museum under the arch. Large, heavy wooden boxes made to haul personal belongings. It took two strong men to lift them and hoist them on and off the wagon, ship, and train.

Years ago, I brought with me such a heavy box which belonged to the family of my grandmother on my father's side. The lid had three faded white letters painted on black. The letters are H.W.E. I guess the letters stand for my grandmother's name, Hermina Wilhelmina Eigeman. That was her maiden name.

No one in the family really wanted that large box and it was standing in the way, so I said that I would take it and bring it with me to the U.S.A. If I had not done that, sooner or later it would have been forgotten, chopped up into dry kindling wood, or just dumped--something I just could not imagine.

When I look at the box and read the letters, I recall my grandmother--"Oma," we called her. She was an old-fashioned Oma, always dressed in black even though she was quite a cheerful person.

There are several wooden boxes, both small and large, throughout our house. Downstairs there is one with the name of the ship on which we sailed for Australia. That one is part of my life. I have a small one in my study which was my brother's, in which he stored the things close to his heart. Right now, I have a cow skull set on top of it and a big rock.

Another box in my study is a large wooden one measuring 24 inches high by 25 inches deep and 38 inches long. It is about the same size as my grandmother's wooden chest. However, this one is built extremely sturdy and I am sure it belonged to an immigrant. Years ago, my daughter found it in an antique store in Riley and even though I bought it, by right it is hers. Right now, I am just using it. I am attached to it because it holds a story I do not know.

Whoever built it was a craftsman as it is very well put together. I am nearly sure the man who built it was a carpenter and what I refer to as a drawer on top, which you can lift out, seems to have been made to store carpenter tools. The corners are rounded and covered with tin to prevent splitting. The handles are heavy duty and you can lift the box. But it is heavy. It was painted black, which is now fading.

In my mind, I see the man sorting the planks to make this rough but sturdy chest. Maybe as a carpenter, he had already built several for other people who were going to the new country of which they had heard so much. Maybe one morning after he had delivered another chest to someone in town, he came back to his shop, saw some remaining planks, and decided to make his own. None of this I have any proof of, all just thoughts. Yet, I do have the wooden chest and it seems to speak to me.

The fact that my daughter found it in Riley makes me think it could have been brought over by one of the Swedes who settled close by. Might it have come from Wales, England? It is all guessing as there are no initials anywhere on this wooden chest. Yet when I look at it, I see a story--not a tale--but a story. I see it packed carefully with only the essentials, as weight did matter. I see it stand near the door the last night the owner slept at home. The next morning, the chest is loaded on a cart and hauled to the wharf. How long that trip took I do not know. Did he wave to his neighbors as he drove by? Did he think about a woman as he drove on, passing the house where he shared coffee, had looked into her eyes as she had said, "No, I can't go with you?" I don't know.

As the harbor and land slowly fell away in the morning mist, the sails filling with the fresh breeze, the man looked back, pulled his cap down, and put his hands deep into his pockets. Overhead the gulls swept by and screeched, while down deep in the ship, the black chest stood stacked securely with other cargo.

The black chest held the man's belongings, but not his life. His life he held in his hands and his head. His decision made, he now was looking forward and as if to hurry along, he walked to the fore of the ship and stood there for a long time, salty spray-water drenching his coat. He loved it.

Someday my daughter will have this wooden chest, but please not yet!

April 21, 2010

The Lonely Swing

A hard south wind blew the new blossoms of the apple tree and the pear trees in front of the house. Other trees were blowing in the wind, shaking their branches. It even seemed too windy for the carpenter bees to land. The wind tossed them about. It was a typical Kansas spring day, partly cloudy with warm sun and a blowing, south wind.

On the north side of the house, the wind blew less and I watched the red swing move slowly in its lee. The swing was empty as if waiting, asking, "Where are you, aren't you coming to play with me? Swing me wild and high in the game of 'got you' when you are not sitting on me. Remember, you would pull the rope and toss me in the air and giggle, and I would try to knock you off your feet when coming down, tied to my new rope.

Oh, yes, I had been lying on the shelf for quite a while. Actually, I was hiding behind a box but then a hand got a hold of me and I was cleaned up, even glued where there was a small crack. A new rope was pulled through the hole and a big knot tied so I would be safe to play with, to sit on, and swing.

The old man who pulled me from behind the box walked around the yard holding me, appraising several branches for the very best spot to hang me. He looked but then chose the one branch of the Bur oak he had planted more than 25 years ago. Might even have been 32 years ago. Thirty-two years is a long time. The branch I was tied to was a sturdy branch that would slightly bend down. I noticed that I reached close to the ground so that a three-year-old could climb on me, especially a sturdy three-year-old with lots of laughter.

It was the first time the young oak tree was trusted to carry a child for play. It is a big responsibility for any tree, but especially a young tree which does not have the experiences the mighty big oaks have with old tires tied to a rope, swinging from their big branches.

But hey, I am not an old tire. I am a true swing seat. I miss the small blond girl with blue eyes. I am swinging here kind of forlorn--just hanging around, doing nothing. I'm bored. Swings should swing. If only she would walk by on her way to feed the horses with Martien or Baba and give me a push, then laugh, and quickly run away so I couldn't touch her. If only she walked by.

Right at my feet, I can hear the basketball sigh. It too is lonely. There was that moment the little three-year-old girl asked for a ball to play with. The trouble is, there are no balls anymore in this household. But Baba suddenly said, 'Yes, we have one,' and out of the box with old newspaper to be recycled, he grabbed a basketball. True, it needed air but there was enough air in it to give it a good kick and it rolled on the grass. Another kick, more laughter, and away the ball went.

I, the red swing, am tied to a branch. Oh, yes, I can swing to and fro but I cannot go where the ball can go. I can't even hide under the bushes. But now the ball is lying near me just as bored as I am. I heard it say, 'I might as well go back to the big box with newspaper. There I can hide and snooze.' It is sad. We are here doing nothing.

But I must say, we do hold the memories. We had fun. It was a great weekend. Actually, between the basketball and me, we will hold the memories and look forward to the next visit. We will be ready and waiting.

The three-year-old and Mummy looked at a snake slithering through the grass. Mummy held her tight then. They were not afraid but asked Baba, 'What snake is that?'

Baba said, 'A bull snake,' and they watched it go, lifting its head every now and then.

It was fun to feed the horses--more fun to groom Poptart. When Martien was small, he rode Poptart. They are still buddies even though Martien is now way too big and heavy to even just sit on Poptart.

From where I hang, I could watch them walk down the paddock and put a halter on Poptart and groom her. Hairs were flying. It had been a while, for Poptart to have a three-year-old around her. But from what I could see, Poptart behaved well. Martien gave the three-year-old girl a riding lesson. Her mummy held her just to be sure. But the little girl even did a turnaround while sitting on Poptart. At one time, she was facing the tail end and grooming the rump with her special brush. I think Poptart loved the attention just like me when I was swinging.

Later I heard more laughter. I could not see what was going on, but I heard that Martien went down to his old room and found his stick horse which came all the way from Europe. His sister had brought it back for him when he was little, all the way from the Old Country. It looked good; well used, but nothing that a good grooming could not fix. That stick horse knew immediately what to do. Martien gave a lesson and showed the little girl how to hold the reins and control the horse. In no time they galloped off.

That weekend there was so much laughter and happiness and that is what we are missing right now. But you know what I heard the basketball say? 'I need a good kick.' I know they will come back. We will just wait together."

May 24, 1989

The Wristwatch

It had stopped running and for a moment I thought, "That cannot be."

When they gave the wristwatch to me, I of course accepted it but did not start to wear it. Letting it slip into my pocket, making sure that I would not lose it. Later, I put the watch in the small drawer of my desk where it kept ticking time, just as it had been doing when my dad wore it on his slender wrist until his death, a quiet death. He died in his sleep while his watch kept ticking, keeping time. It is a strange thought that time did not stop when his heart stopped beating. Of course, it never does when someone dies, and as a matter of fact, time heals.

His death was not a sad death. Actually, his last years were saddened by the earlier loss of two of his three children and later his wife, my mother. Somehow, he changed, he missed her too much. His whole life he had helped and encouraged other people, but he found it difficult to encourage himself, and he wanted to die. He had done all his living. I went home as often as possible, sometimes more than once a year. Brief visits, a week or ten days long, but I know they were highlights, especially the last few years. I would write to him the date I would arrive in Amsterdam by plane and I would tell him to expect me for the tea in the afternoon. The last few years, he lived in a rest home and I knew when tea or coffee was served.

Arriving in the afternoon allowed me to get over jet lag and pick up a rental car. I would slowly drive to my Dad, enjoying the familiar scenery of the Dutch landscape. I would pass scenes I remembered from my childhood: main streets of small towns, church spires, rivers with the boats, trees in pastures, hedgerows, farmhouses, the red and white cattle, the house where we all grew up.

I would make sure I arrived on time, early enough to park the car and walk to the home, ring the bell, and enter. Dad would be sitting in his chair. He would slowly turn his wrist, look at his watch, and say, "You are here." For a while, he would not say anything but just savor the moment as he took another look at his watch, to celebrate the moment.

After a while, I would say, "Let's go for a walk," and in the hall, he would reach for his coat and hat and we would step outside. We slowly walked the paths through the park and finished up at a favorite spot, a bench underneath a large oak tree, and there we would sit and talk. Talking would go slowly but it need not go fast because for that moment we had all the time in the world. Every now and then, he would slowly turn his wrist and watch the time.

That place under the large oak became a special place and I can still imagine it. I see it when I close my eyes. I wonder how many times Dad went down there to quietly sit, read my weekly letter to him, and watch the time. We asked him to come live with us but he did not want to. He was independent.

When the old Timex I wore finally gave up and quit running, I took Dad's watch out of the drawer and carefully put it on my wrist. I used a different hole on the strap. And as it had been doing all along, it just kept running, keeping accurate time. That was until it stopped a few days ago. Luckily, it only needed a new battery.

It is 4:34 in the afternoon now and as I look at the watch, I see a bench under the big oak tree, not an empty bench, but a bench full of memories. I hope that wherever you live you have such a "memory place." If not, you better create one. I know I have for my children and it too is a happy place--here in Kansas.

June 6, 2007

D-Day

Today is June 6, 2007, and if we know our recent history, then we remember that 63 years ago today, General Eisenhower made the crucial decision to proceed with the invasion of June 6, 1944, on the French coast. Among the designated beaches, Omaha Beach was where American troops landed and where there was so much carnage.

But just like the landings on the other beaches, the soldiers who survived, prevailed and pushed inland. They crossed the beach, climbed the dunes or steep shore, and neutralized the enemy fire. It was done at a tremendous cost of young lives at sea, in the air, and especially on the ground--on the beaches. That first day, 10,000 plus allied soldiers died, of which 6,603 were American casualties.

I just finished reading once more, *D-Day - June 6, 1944, The Climactic Battle of World War II* by Stephen Ambrose. I have read it before and know I will pick it up again and read it or reread parts of it. I will look at the photographs and the maps, but especially the photographs of the young faces thrown boldly into this hell. And as I look, I softly say, "Thank you for what you did."

For what they did was the beginning of the end of the German occupation of Europe. The war would last nearly another year for us, but the Allies had landed and established beachheads. Some units moved further inland that first day, then others depending on the resistance thrown at the allied forces.

I remember how everyone tried to listen to the radio--not allowed. But I still have a vision of my father standing, listening, and saying, "They have taken Saint-Lô on the Vire river." The town of Saint-Lô is well behind the landing on the beaches, an important town on the march inland. By the time they had taken Saint-Lô, the Mulberry harbors were in place and war material poured into the country as well as more and more troops.

But the book *D-Day* talks about what Cornelius Ryan called "The Longest Day," a title of another book about the landings. Reading about the development of that day, you know it was an awfully long day. For many, the last day of their young lives. The book *D-Day* is full of anecdotes, happenings written down as told by the survivors.

A soldier wrote in his diary that on the evening of June 6, before bedding down at Sainte-Marie-du-Mont, he thanked God for helping him to live through the day and asked for his help on D-Day plus one, June 7. Making a promise to himself, if he lived through the war, he was going to find an isolated farm somewhere and spend the remainder of his life in peace and quiet.

It says that in 1951, this soldier got his farm in south-central Pennsylvania. By now he, like many of the others, will have aged. And to some of our listeners who participated that day in the landings on the beaches of France, I say and repeat to you, "Thank you for what you did."

I have visited the Eisenhower Museum in Abilene several times. It is somewhat strange that this man who meant much to our lives lies buried in this small Kansas town. Whenever we have visitors over from the Netherlands, we have gone for a visit to this museum as well as on special occasions when there was a particular remembrance.

Whenever I walk the displays, passing the mock-up of the landing craft with the ramp down, I pause. I try to visualize, realize, what it was to step out, fall in the water, try to get up, crawl, wade forward, and be shot at. The fact is, it happened 63 years ago today and for those who survived, it is a whole lifetime gone by. A strange thought, it really does not seem that long ago and yet it is.

An older friend of mine, Ron, who was part of the invasion forces was wounded and taken back to England as were many of the wounded. His whole life he has carried some shrapnel in his body, besides suffering from deafness due to exploding shells. He had a very full life--a great man.

If you are a veteran of the D-Day invasion, whatever part you took--Navy, Air Force, paratroopers, or infantry--you will have your own thoughts and memories of that day. The hours that a whole nation of those left behind anxiously waited for news from the beaches and the battlefields about their sons, husbands, fathers, brothers, sisters, and uncles.

I am very happy that Ike was a son from Kansas, raised in the town of Abilene. I must go back and pay a visit to the museum. But it is today, June 6, 63 years ago, that they jumped off the landing crafts and waded ashore, drove their tanks and jeeps off the ships, many of which sank the moment they left the ramps. Read the book--remember the history.

July 23, 2008

His Hands are Like Grandpa's

She looked at my hand, turned to her mother, and said, "His hand is like Grandpa's hand."

Actually, I think she had looked at my thumb and seen the large black-and-blue blood blister which is slowly moving to the outer edge of the nail. Right now, it is in the center of the nail. It is a large blue-black spot slowly moving along. Like an island drifting in the ocean, a small ocean.

When they noticed it, the blue-black spot had just appeared along the upper edge of the thumbnail. It was the result of missing a nail and hitting my thumb. A good solid hard hit--ouch! When it happened, I turned around on my heels, held my thumb up, and tried very hard to think of other things than the swelling pain and pulsing thumb. When you are right-handed, it is always the left one which gets the blow--ouch!

I don't know if I said anything. I don't think I did as there was no one around to listen to my exclamations. Sometimes they are in English and other times, Dutch. Which explains why my children and now grandchildren know and can say some keen Dutch words, words which are spoken when such a painful event takes place. It is one way to pass on a language. Although, I am sure my wife does not approve.

If I remember right, I just kept on working after the initial throbbing went away. I have always found it is the best thing to do, just like when you twist your ankle. You just keep walking and ignore it. So, I kept hammering nails and building the needed cavalettis--the jumps. But I used the hammer a bit more carefully which actually made me miss more often. Ultimately, I forgot the blow until it changed colors and ever so slowly, over the next few weeks, showed itself under the full nail and was noticeable.

Old hands can be very interesting. I have all my fingers. My hands look like my father's hands, only stronger, suntanned, and with a few age spots. Old hands can be deformed from accidents. Mine are not, as I generally treat them well. I only have two and appreciate them for what they do for me and allow me to do. They are calloused enough that I don't blister.

Until a few years ago, I never wore gloves--not even when fencing. I remember one time fixing a water gap that had washed out between me and my neighbor. As I was pulling wire and twisting and winding it to close the gap, working without gloves, my hands and arms had been scratched by the jumping wire. I was bleeding on both hands and arms. It must have been bad--a bad enough sight--as I later heard from an acquaintance that had been told about it by the neighbor. "Here he was standing in the dry creek bed, pulling the wires and bleeding like mad." By that time, I only had scars to show. But I had closed the fence. I think that it was after that when I started using gloves.

Another reason was because I was given some work gloves under the Christmas tree. And once you start being soft, it is the beginning of the end. I now have two pairs of gloves lying on the seat of my pickup and notice, I do wear them. Working with a new roll of barbed wire, I have had it happen that my thumb got caught in a jumping barb and as the wire unrolled, it twisted, trying to take my thumb with it. The last time that happened, I was glad to have gloves on, knowing how it rips out a piece of flesh when not protected.

Old hands can tell stories--steady hands, trembling hands, suntanned hands. When the young girl softly said to her mom, "Look, his hands look like Grandpa's," I felt happiness. I was holding a horse by the reigns and could imagine her grandpa doing the same. I could see her grandpa moving hay bales, fixing a gate, lifting the foot of a horse, cleaning the mud, and looking for stones. I felt I knew her grandpa, even though we never met. I could see him cleaning a stable, quickly moving a loaded wheelbarrow. I could see him looking at his granddaughter and loving her like I love my granddaughters and grandsons.

The blue-black spot under my thumbnail is only a quarter inch by a quarter inch. Right now, it is in the middle. It is more than eight weeks ago that I hit the thumb--ouch! I'll watch it slowly move to the edge and then I will have to cut it when I cut my fingernails. I'll miss it when it is all gone. It is a reminder of a very happy time--working, watching my son ride and jump. Maybe that is how come I hit my thumb. That's it! I was watching him riding, training, and clearing the jumps--ouch!

You know what? I would gladly hit my thumb again.

October 28, 1998

Homecoming

To be able to come home, you have to leave first.

Throughout my life, I have left home many times. When I left the Netherlands in October 1952 to go to Australia, I had no clue when I would return home--that is, to the Netherlands--if ever. Those years, shortly after World War II, people traveled by boat. Most passenger ships were refurbished liberty ships that had survived the war. Traveling by ship was a slow way of getting from here to there. It took five to six weeks from Amsterdam to Sydney, Australia. It was a five-week trip through the Suez Canal or six weeks through the Panama Canal. I have gone both routes--the Suez Canal voyage twice.

When I came to the U.S.A., the boat trip would take six days. It was a tremendous improvement over the long voyage to and from Australia. It was one of the reasons why we decided to come to the United States--being closer to Europe would enable us to stay in closer contact with our families back home.

Traveling by boat is a real home going and homecoming experience. The home going is a slow process--the ship plods along mile after mile after mile. It keeps going day, night, through storms, and rough seas. The latter will slow it down, but she keeps plodding, rising up and down, crashing into the waves, or smoothly steaming along. When entering the final harbor, that is the homecoming. You stand at the railing and you see all the activity--the tugboats pushing, pulling, and shoving. You can wave to family and friends and sometimes shout greetings.

As the years have gone by and our family grew up, we have been able to keep in contact with the extended family in the Netherlands--only now we fly. The flight from Detroit is a mere seven hours. Family regularly come to the United States and sometimes we hear that they have been in the country to visit other relatives or simply on vacation, but they did not come close to us, so we did not see them. Sometimes they are very close but don't realize it.

A few years ago, my nephew, Mark, was in New York. He drove from New York to Montreal for a weekend visit and as he passed through Albany, he was less than 25 minutes from his cousin, my daughter, who lives with her family in Saratoga Springs, New York. Just think about it. This would not have happened 40 years ago. But now family is scattered everywhere. We call each other, tell them to stop by when they are close--we have coffee ready and there is always a made-up bed. When people stop by, we make them feel at home.

Sometimes, as I am getting older, I think about the meaning of "being home"--"coming home." I think about all the places we have lived. A few years ago, we visited friends in Blacksburg, Virginia. As we sat and visited with several of our friends, they told us, "It is time to come home," meaning we want you folks to come back to Blacksburg. When friends tell you that and mean it, it is a beautiful experience. But after 24 years in Kansas and raising our family here, Kansas is home.

Flying into Kansas City and seeing my daughter and granddaughter waiting for me at the gate is coming home. Calling your wife when arriving in Manhattan to tell her, "I got in on an earlier flight, could you come and pick me up?"--and when there is no one home you call a friend and your wife is there drinking coffee--that is home. You are returning to a place you know and where people love you. That is home!

I always like to make a homecoming special, if not for myself, then for others. Sometimes you have to start that while they are young. I'll share a moment with you. A few weeks ago, I was in Portland, Oregon. On Saturday, I had some time to myself and I rode the trolley into town. Right in the old part of town, there was the weekend craft fair. The trolley went right through the middle of the market, slowing down and ringing its bell. I decided to step off and walk the row of tents.

Moving along, I came to a jewelry stand where custom made jewelry was on display and for sale. There were some neat pieces made from dinner table forks, spoons, etc. For instance, rings were made from spoon and fork handles. The handles with their decorations were cut off and simply rolled into a ring--silver, stainless steel, etc. I looked over the tray with rings and picked one I thought my daughter would like. I paid for it, received it, and tucked it away in a tiny bag as if it was an expensive jewel.

Flying home, I had a thought, a plan, knowing that my daughter and six-year-old granddaughter would be at the Kansas City airport. After a big hug, I gave my granddaughter the small bag with the ring in it and told her, "You hang onto this, don't lose it." I watched her tiny hand hold the small bag tightly.

When we got to the gate for my next flight, which would take me to Manhattan, I said to Arianna, "Do you still have the bag? Okay, now listen to Baba. What you hold in your hand you now give to Mama, but when you are grown up, it is yours."

Arianna then gave the tiny bag to her mom and as my daughter retrieved the ring, Arianna's eyes got bigger and bigger. I repeated what I had said earlier. Soon I had to board my plane and was gone.

Later that evening, I called my daughter to make sure that Arianna had understood my gift. My daughter said, "Papa, don't worry. She did." Then she told me that while driving home from the airport, six-year-old Arianna had turned to her and said, "Mama, when will I be grown up?"

November 9, 1994

Good Firewood

When I cut firewood, I want to use all the wood--not waste any.

This year, on several occasions, commemorations have taken place to remember what happened 50 years ago during WW II. Earlier this past summer, we remembered the landings at Normandy and other beaches on the coast of France. This fall, the airborne troop landings and battle near Arnhem, the Netherlands, was remembered. I lived about twenty miles away from where it happened.

The movie *A Bridge Too Far* was based on that event. That was also 50 years ago. After that battle was lost, I remember the German trucks passing through our town. The open trucks had standing room only for the allied soldiers taken prisoner. I remember how amazingly cheerful the soldiers were, giving the V sign. I was twelve years old. I can still see it all. It is like a photo in my mind.

That following winter, when the war dragged on, it was very hard for many people, especially those who lived in the cities. Often with no water, no heat, no coals, no wood. Those who lived in the countryside would look for dead and fallen wood, but because everyone was looking, firewood was scarce. I remember we would go out with handbags and bicycles without tires and load them up like donkeys with sticks and sprigs of wood, pinecones. We were never cold, but it was a lot of work to find, cut, and stack the wood, being small wood such as branches and twigs.

At this time of the year, I have a chance to remember those activities as my son and I cut firewood on the farm. We cut firewood from large dead trees or trees which are leaning. There is plenty to choose from and it does not take long to cut a pickup load with hard ash or oak, even walnut. The larger wood has to be split and the heavy maul makes it an easy job, especially if my son does the splitting.

What was very interesting for me was my conservative attitude toward the trees. There was the trunk to be cut as close to the ground as feasible. Then all the branches--and I mean all of the branches. My son's chainsaw is used to lop off the tops of branches fairly close to the trunk, leaving a lot of brush--good brush with good wood. It is the brush that reminds me of my parents and us children searching for twigs and sprigs to carefully burn in the old, wood stove.

When I am working with my son, we get more wood from a tree but I know he is not really interested in the smaller stuff. He needs the big stuff that needs splitting or branches that are big and heavy which when cut, still provide heavy wood.

But it is not only the memory of 50 years ago. I know that today some people in Pakistan, India, and many other parts of the world where forests have been depleted, have to go a long way to collect wood on which to cook. And here I am throwing branches aside because they are not thick enough. Oh, but I just don't throw them to the side. I stack them to make a nice brush pile for the rabbits and other critters. So, I am not wasting the wood, but I am surely not burning it.

I have never told my children how my parents, their Oma and Opa, went out to collect firewood during the last year of the war. It was the same for my wife's parents. She, as an eight-year-old, and her brother would follow the German lorries which hauled coal for the Germans. Sometimes a desperate and courageous person would try to push a bag of coal over so it would spill and people would pick up the few coals which fell onto the street. Your pockets full of pieces of coal were a gift from heaven. They lived in the city and there were no more trees to cut down.

Today, 50 years later, I still can feel excited when I pick up a heavy branch that fell from a tree and I can judge it will make a few good logs of firewood. It is 50 years ago and I know it is crazy that I still respond that way. I think I always will. But I have to tell my son so he will know why I need to cut up the smaller branches--it is all good firewood!

January 11, 2012

The Gift of "Poop"

This story has an aroma to it and is somewhat nostalgic.

It started with a gift my grandson gave me for Christmas. As time flies, that is already more than two weeks ago. The gift was placed in a card and upon opening the envelope and reading the card, it said, "I invested in poop power for you." Right above the words was a drawing of a pile of whatever--could have been elephant dung. On the inside of the card is the explanation. "In your honor, farmers are being trained to transform manure into agricultural wealth." It is part of Oxfam America's gift program that gives in two ways: providing a symbolic gift, as well as supporting people in need to overcome poverty.

This year, other gifts were given--bees, chickens, and goats in honor of us. It is a happy feeling to know that someone receives a gift that will make a difference. But I admit I had never yet seen or heard of the gift of "poop" or I should say--*manure.* But I like it. To grow crops, one needs to build up the organic matter and fertility of the soil. And the one way of doing that is by collecting, storing properly, and distributing the good stuff given back by farm animals. Having acknowledged the great gift, I wrote my grandson a letter.

I told him that as a teenager in the mid-forties, one of the winter chores during vacation was to keep the stalls of the dairy cows clean. Manure had to be removed and fresh clean straw put out. Then cows were tied to stanchions, whereas now they have large covered, loose stalls.

Twice a day, we moved behind the cows and scooped, taking the manure to a storage place. From there, the collected and stored manure was loaded on a horse-drawn, two-wheeled cart. When the cart was filled to the top, the load was driven to the field where crops were grown. Sometimes those fields were close but often, further away. When the latter was the case, an old sack was thrown on top to make a soft seat to ride to the field.

As you can imagine, doing that kind of job required old clothes and a strong tolerant nose toward the good stuff you were hauling--it represented money in the bank, so to speak. You wanted to do this chore when the ground was frozen so the horse and cart had it relatively easy.

At the field, you removed the backdrop of the cart which was not always easy and often required some prying. Once removed, a hooked fork would pull manure down in small heaps and while moving the horse and cart forward, more heaps were dumped until the cart was empty, tipped, and scraped out.

On the way back, you rode in the empty cart. You had to make sure that the hook to hold the box in place was re-attached. If not, by moving backward while riding you could tip the cart with your weight. I can tell you, now and then it happened and a young horse would bolt. But they always ran for home.

Sometimes on the way back, after a good clean scrape, sugar beets would be loaded so as not to waste the trip. The beets would have been dug in along the edge of the field and protected from frost similar to how potatoes were stored. It was all very simple, time-consuming, and steady, hard work. But I enjoyed being a part of it until school started again.

High above the fireplace in our home hangs an old Dutch painting of a winter landscape, a lane with big trees and snow. Coming toward you is a loaded horse-drawn cart with a man walking beside it. I love the old painting, as I know the scene. It is where I grew up and worked as a teenager. But one day something happened. I noticed a light reflection in the pink evening sky of the painting. Suddenly, it was alive. As the light slowly moved over the painting, it put the focus on the horse and cart and the path it followed. It was very special; I saw details I had not seen before.

I looked around and then saw where the light source came from. It was the sunshine reflecting on the glossy cover of a book lying next to me. The sunlight itself would never shine on the painting. But at that one moment, the reflection from the glossy cover was thrown just right--amazing. Also amazing was how much life it brought to the old scene--the sky, the trees heavy with snow, the muddy track, the horse, the cart, and the old man.

The strange thing was that as I wrote my grandson about my involvement with good old poop and remembering my earlier experience with the painting, I suddenly smelled it again. Here I was sitting behind my computer and I smelled manure. It was a clean, warm, fresh smell. It took me back to the dark days of winter and the job at hand of scooping, hauling, and spreading manure while walking beside a horse-drawn cart.

Sometimes we get a gift which makes us happy for different reasons. My gift became a personal memory for which I am very happy!

Sense of Place

Memory of the Old Country.

February 27, 2008

Chased by Light

She wrote on the card, "Light is to me as location, location is to you!" What she meant by "location" is a deep sense of place. And she is right. A sense of place is important to me and has been all my life. I have tried hard to create it for myself and for my family--the children. Looking back, I think I have been successful. Although I know at times, it came at a price when we had to move away from where they, at that moment, felt comfortable surrounded by their friends and familiar places. They felt secure.

It was the same for my wife. But as we sat last Thursday morning enjoying a cup of coffee and a quiet moment in a warm, cozy house while cold snow came slowly drifting down and a yearling doe stepped out of the trees to enjoy the cracked corn I had placed in the trough for her and the others, my wife said, "What a beautiful place this is!" It surely has become that. I looked out and saw the thin coating of yet another snowfall. I saw the white, the light of the snow, stand out against the dark and shadowed cedars. It makes for deep intimacy in the trees where the deer withdraw. I saw the young doe show its white flag and prance away.

To see light it helps to be in the dark. We all know the expression and experience of seeing the light at the end of the tunnel. Here in Kansas, it is possible to step out of shadowed woods onto the open and light prairie. I know that experience from our farm. We have a road which on a summer day is heavily shaded by oaks and other trees. As you walk toward the west, you step out onto the open prairie meadow we cut for hay. I like that walk from dark to light. I also like the reverse--walking from the open grassland into the woods, especially on hot summer days.

A week ago, the moon went hiding behind the shadow of the earth. I did not see it. I went to bed. But in my mind, I could see the natural drama unfold. How the earth moved between the sun and the moon and the moon would be dark except for the ring of light around it during the full eclipse. My mind could see it all, knowing what was going to happen. I thought about earlier people who did not know or could not explain--except through mystics--the end of the earth, their world. And then slowly, the moon would come back as the earth revolved. You could, in your imagination, hear the drumming and see the dancing. "We did it! We did it! She is coming--coming back."

Once in my life, I observed a low flying swan--a black swan--flying in front of a full, low hanging moon. That was in Australia. I still see the momentary image of this big bird moving in front of this huge white moon. Coming from Europe with its days of overcast and rain, the Australian light was a totally new experience. It seemed that light was all around you, all the time. The eucalyptus with its narrow gray leaves reflected a shimmering light. It was different; a near-white light.

A few days ago, I took a walk outside. There were no shadows because there was no sun. To experience light, one can create gardens that show light simply by creating dark places from which one steps or looks into or to the light. To create such a place, one has to think about the effects. Just like when I walk on the farm from the shade into the light of the prairie.

We can also enjoy and celebrate the moon and its light. To celebrate the moon even more intimately, we can create a garden, or a place in the garden, where a small pond reflects the moon or where white flowers catch the soft light enhanced by fragrance. There are many ways to enjoy light and celebrate it. Glossy leaves or gray foliage all enhance moonlight. Light is a reflection on snow or on water where it mirrors in the rippling waves as if thousands of crystals are floating.

For those who celebrate light, there is a challenge to design a house and/or garden, to create a home--your place--seeing, using, feeling, celebrating light, touching color. It was my daughter who gave me the book, *Chased by the Light* by Jim Brandenburg, and who said, "Light is to me as location is to you." I think I understand.

February 20, 2002

The Old Farmstead--Restored

They fell in love with the property; a unique old stone house, a couple of barns, some good bottomland along the creek backed up by the grassy hills. Thousands of years ago, the creek cut its swath through these hills. The couple of hundred acres is no longer a viable farm unless one grew specialty crops in the rich bottomland. The hills have been overgrazed and woody plant material has invaded what was once upland grazing.

Mind you, there can still be good grass again, especially if the hills are rested and the torch is put to the dried grass. If the new owners burn for several years in a row, I expect the rough-leaved dogwood will recede and thicker stands of native grasses will thrive. But as I said, except for the bottomland, the farm cannot be really productive as is the case with so many small farms struggling these days.

Right through the center of the farm is a glen. The naturally eroded draw gently runs from the top of the hills into the creek. The area is wooded with oak, hedge, Kentucky coffee trees, hackberry, elm, locust, and of course, cedars. Walking up or down the glen on this natural wooded trail, you have no feeling of the wide-open Flint Hills on top. Walking through the glen, you are enclosed and the wind can't get you. It is a beautiful habitat for wildlife and there are plenty of deer and turkey tracks in the soft ground. The last remaining snow is hidden in cool pockets of a boulder or a fallen tree stump.

No, they did not buy the place to farm. They wanted it for its "sense of place;" its history chiseled into the stone above the door. The roofline of the two-story, old country house is still straight. Soon there will be additions to the old house. The drawings worked out by the architect are sensitive and respectful of the older structure, yet the planned addition will make the old home very livable as a remodeled country home.

Where the old home had little need to look out over the scenic valley, the new addition will. As a matter of fact, it will highlight the long and wide vista to the south from large windows and a patio. Protecting the patio will be an arbor over which vines will climb, hanging baskets will hang. The remodeled old, stone house will be a home for the new owners.

The farmer who worked out in the fields the whole day had little need for a patio and shaded arbor. He had walked the open fields all day, first behind his horses and later sitting on the tractor. The rusted farm machinery stored under trees and remnants of tack hanging in the old barns speak of earlier times. The date in the stone over the door reminds one that the house is more than a hundred years old.

The old barns with antlers hanging on the outside will be restored--doors hung on hinges again. Buildings will be cleaned out, junk hauled off. It is amazing what farms collect. I have marked trees that must come out as they clutter the yard and are a danger for man or beast. Walking up to a vicious locust tree with mean-looking thorns, I accidentally stepped on a broken branch and impaled my foot on a nasty thorn. It went right through the leather into my foot--what a tree--out!

In the right place, away from the yard, I could have left it as a safe nesting tree in which no cat would climb nor snake would crawl to rob a nest. I carefully selected young Bur oaks growing randomly on the property to replace, in time, the huge, big oaks around the farmstead. They are beautiful trees with wide-spreading branches providing shade, but above all, character to the farmyard and farm landscape.

It is the natural beauty and the culture--agriculture--brought to this valley and especially to this acreage which made the new owners fall in love with it. Now they are going to put their stamp on it. Their stamp will not be one of growing crops and harvesting, even though that will continue on the rich bottomland. As time goes by, their harvest will be silence--privacy--nature--the moonlight lighting up the wide-open field framed by trees and hills on both sides and dissected by the stream.

If a hay crop like alfalfa is grown, then the scent of curing hay will come into the house on the south wind. And after the hay is baled and hauled off, the turkeys will flock along the edges to pick the grasshoppers. From the hills, the laughter of the coyotes will echo across. It is a beautiful place and I am so glad that this place, this old house, these old barns, valuable for their authenticity and history, will have a new lease on life and will continue to grow older together with the mighty oaks standing in the yard.

April 15, 1998

Kaw Valley Heritage Alliance

Tuesday, a week ago, I lay awake in bed at daybreak when I heard the commuter airplane take off at the Manhattan airport and climb into the gray morning sky; another day of low hanging clouds.

But I was not thinking about the weather that morning. The thoughts I had were about the talk I had to give later that day at the Kaw Valley Heritage Alliance's Third Annual Conference in Topeka. The conference was to be held at the Kansas State Historical Society Center just off I-70. The theme for this year's conference was "Caring for the Kaw."

As I listened to the airplane gain altitude, I mentally followed the airplane on its flight to the Kansas City airport. It would roughly follow the Kansas River, also known as the Kaw, through the valley, more or less.

On takeoff, I would have seen the University--Kansas State University that is--disappear underneath me. Looking back, I would have seen Tuttle Lake and the dam. Flying east, I would have seen all the familiar landmarks: the small towns, the tributaries of small streams emptying into the Kaw.

Listening to the droning of the plane fading away in the distance, I thought next of being a raindrop falling on our farm in Pottawatomie County. How that raindrop would fall on the Flint Hills, following a cow path down to the pond. In the pond, it would take some time for that drop to come out of the overflow pipe and continue its way down the draw and empty into McIntyre Creek. Then it would flow on over the crossings, under the bridge, and past my neighbors. It would follow the valley floor of the creek with the Flint Hills on both sides. The Big Blue River would bring the droplet to the Kansas River and then it would flow east into the Missouri River. It intrigues me that our farm is part of the headwaters of the Kaw Valley watershed.

Twenty-four years ago, I could not have lain in my bed and seen all these connections. It is all part of having a sense of place--knowing who you are and where you are--where you belong.

I could not hear the plane anymore but was quite sure where the plane was. I followed its flight and imagined myself sitting in the plane and looking out the window.

The greening of the landscape was all around us now. The wheat fields were green, but so were the alfalfa fields with their first flush of growth. The farm ponds were filled to the brim, many running over. The full ponds looked like mirrors held up to the sky. Towns like Wamego, Rossville, Lecompton, and Tonganoxie would pass underneath, or at least be, more or less, below the air route taken. I know the towns, small towns, the cafes, and I know the roads--the landscape.

A few days ago, I received a letter and the 32-cent stamp was a commemorative stamp of Iowa. It had the year 1846 printed on it, the year of its admission as a state. But what intrigued me more was the picture on the small stamp, no bigger than an inch by an inch and a half.

On that picture was, more or less, a 400-acre farm with a cornfield running down to a road. A road which ran down into a valley with a farmhouse, a forest. Trees grew behind the house, following what I supposed would be a stream. In the distance, the trees flowed into the grasslands on the hills--it all melted into the blue sky of which there only was about a quarter-inch strip left along the top. The small stamp showed much landscape and the culture in agriculture.

I wished we had some similar postage stamps of Kansas scenery. The practical joker would say that is easy, just draw a line halfway through a stamp, color the bottom half brown, an earth tone, and the top half blue and you have a Kansas scene.

Okay. That could be western Kansas, but allow me to move the horizon line up or down one third, two thirds, or even one-quarter sky and three-quarters earth. Or maybe reverse it and add some different colors to the sky. It could be abstract but yes, it would be a Kansas scene and a Kansas postage stamp. Another idea would be to show the Flint Hills or the Kaw Valley from a bird's eye view of the river flowing east through the hills through rich farmland--a Kansas landscape.

June 23, 1999

I am of Kansas

I am not a linguist, but I love the feeling of language--what do words say? What message or information or emotions do words convey?

For Father's Day, I was given the book *History of the Donner Party: A Tragedy of the Sierra* written by C.F. McGlashan in 1880. We all know the story of the Donner Party. How they set out too late and got caught in the winter storms high in the mountains while trying to cross into California. As a matter of fact, I believe the party camped at Alcove Spring near Blue Rapids here in Kansas on their way out.

Even though I know the story, I have started to read it and with the maps placed in front of the book, I can follow the route and travel with that group of people. The maps make it all come to life for me because they take me there.

I have driven the highway pass through the mountains and in a few minutes, you drive what they struggled to climb over for days and weeks. Glancing from the car, you see into the landscape and appreciate its ruggedness, even when seen in the summertime. You know that you only get a glimpse.

You actually should park your car, jump the road barriers, and strike out on foot, trying to find a trail among the boulders and imagine bringing horses and pack animals along. Even on a summer day, one comes to realize that crossing the mountains was a daunting task. It becomes nearly impossible when snowstorms and blizzards move in.

But coming back to the title of the book, *History of the Donner Party: A Tragedy of the Sierra,* I have thought about the word "of"--*a tragedy of the Sierra*. I think that little word "of" makes the tragedy part of the whole Sierra region--all the mountains and valleys, streams, boulders, sky, day and night--all life found there. It makes it all-encompassing.

Whereas had the author instead used the two-letter word "in," then the meaning of the tragedy would have been different. A tragedy "in" the Sierra would have placed the historic event in a particular location which is, of course, what happened. But to me, it would have seemed as if, with a scalpel, we cut a small piece out of the Sierras, taking away the awesomeness of the whole Sierra mountain range, focusing our attention on a very small area--only where the incident happened.

Too often when we look at a landscape feature, we look only "in" the landscape but we do not realize that what we see is "of" the landscape. It can be as simple as a stone, a tree, a ridge, a river, a town, or a city for that matter. Looking only at each item, they are all quickly out of context.

When my oldest son, who had visited us and with his family camped on the farm, finished packing his car to drive back to Raleigh, North Carolina, I gave him a big stone. I had picked it up the day before out of the soybean field. The stone was a big blue and white piece of flint with limestone attached. Two sides were flat and with the weight it would make a great book rest.

Taking it home, I scrubbed it clean and I remember while doing that I was thinking, "Should I or should I not do this? Should I leave the soil on?" The question being: did I find the stone "in" the field, or did I take it "of" the field? There is a difference. The words "in" and "of" give it different meanings. I did not want to give him a stone which I found "in" the field. I wanted him to have a stone "of" the field.

When back home he views the stone with even a casual glance, I want him to see the soybean field, the long straight rows. I want him to see the creek with the bordering trees and the crossing, to see the Flint Hills with the cows grazing on top. I want him to hear the owls calling in the early evening, to hear the waterfall near the spring, the frogs starting up their noisy gawking at dusk. I want him to "see" all the parts of the landscape, the landscape of the Flint Hills of Kansas.

What is possible for a stone is also true for a tree, a river, a road, a town. All you have to do is follow the ends of your thoughts--dare to think big. There is a reward or a price you pay--an emotional price--when you are of the land. It all depends where you are. I am living in Manhattan, but I am of Kansas.

June 21, 2000

Watching TV

On June 16, 2000, *The Manhattan Mercury* had under the editorial *Another View*, a short article titled "Let's get children outdoors, away from television sets." The article bemoaned the fact that too many children are overweight and that instead of inactive video games and computers, children need to participate in bicycling, football, and ballet--just to mention a few activities.

To be honest, I am glad I am not raising children today. Ours were raised before computers were common household tools. As a matter of fact, our youngest son, for a while, grew up without a TV. The old TV had given up after one of the older children gave it a kick and it stopped. I said, "That's enough, I can live without the TV." And we did for quite a while. Until one day my wife realized that our youngest son grew up as a TV illiterate. He did not know who Big Bird was. So, we reinvested and ever since have had a working TV.

Now all the kids are grown and gone and it is the grandkids who run to the basement, pull pillows from the bed, and sit or lay down to watch their favorite programs. However, when they stay with us, after a while my wife will say, "Hey guys that is enough TV, out you go."

After a drink and snack, they'll storm outside and soon they are involved in play--imaginative play. As I mow the yard, I can see the evidence of their playing. We still have a tipi under the cedars. The cover is torn and I should take it off. There is the old fort, now crumbling. It was not built to withstand a major siege or a Kansas winter. One could say the pioneers overwintered there, found shelter, and then moved on. I am sure that was in their minds as they built it and played around it. If I decided to take it apart and cut up the poles for firewood, I would have several evenings of warmth next winter. However, I find the poles much more useful as poles because I know they'll want to build something else.

What makes our yard a haven for children to play are the trees, the paths through the meadow, the hideaway places, and secret corners. Among the trees, there are climbing trees and I can see that they have been used because of the strings tied to the top of some of those branches.

And this brings up the point: with summer vacation, where can your children play--really play? I am talking about small children who can play for hours in a shaded sandbox. On a hot day, add some water and no one will come begging for the TV. But the sandbox has to be a real one--two feet deep and at least six by six feet. I like them even bigger, especially if you want neighborhood kids to come over and play at your place. If you shape the sandbox in an "L" shape, it will give them more corners to do their own thing.

The other day I stepped into a yard where a construction crew was doing a remodeling job. There was a big pile of sand under one of the trees. Near it was a Tonka crane and truck. I loved it. A kid was playing with the sand, using his imagination, making truck sounds, "brrrrrr--brrrrrr." What about a treehouse or hideaway? In the same yard, I noticed that the children had brought their plastic play furniture outside and had made a home under the arching branch of shrubs. They had set up house. Our kids played just the same. In 30 years, little has changed. That is, if given the opportunity and encouraged.

Yards--front, side but especially backyards--might go through transition periods. As the children grow up, they move away from home--from sandboxes to basketball courts--and that is how it should be. However, don't let good years slip away. Start while they are small and create those kinds of play stations which they need at the age they are.

If you grow a vegetable garden, give them their corner to grow things. Supply them with the garden tools they can handle and make sure that those tools are quality. For a child who is seriously preparing his or her garden, to break a puny spade or fork is quite traumatic. Good tools are well worth the money and above all, well worth the memory.

Among my many garden tools, I have my mother's spade and fork. They are just a little smaller than the regular man's garden tools. These tools were hers and we knew not to touch them. They are hanging in the garage well cleaned and oiled. When I look at them, I see my mother working in the garden--good memories.

To come back to the topic of children, especially the small ones, you don't need a big yard to create play areas. All you need to do as a parent or grandparent is switch off the TV and get going--get started.

There are seven things in a yard to stimulate play. I will list them and will add on the adult version of the same thing but now looked at with adult eyes.

1. **Enclosure**-children want to feel safe (Adult-fence)
2. **Entrance**-children like to enter a place (Adult-gate)
3. **Hideaway**-children like hideaways (Adult-gazebo)
4. **Hill**-children love to climb a hill (Adult-boulder or rock garden)
5. **Meadow**-children love to run through or in a meadow (Adult-lawn)
6. **Water**-children love to play with water (Adults-pond)
7. **Tree**-children love to climb trees (Adults-sit in the shade)

Just think of how many things you still have in common with your children. All you have to do is help them get started. Of course, I am talking about children who grow up in the urban environment. On the farm, things are different. The tikes play around the house and barn. Safety is an issue. But once they grow up, they start to help with chores--the fun ones and those not so much fun. But they are part of the farm and quickly learn responsibility.

Friends of mine have a voicemail recorder I really like. If they don't pick up the phone, you hear, "We can't come to the phone right now. We are either chasing cattle or children, but please leave a message and we will call you back."

I love it--"We are chasing cattle or chasing children."

July 23, 1997

Grandkids

They have taken over the house. The dog, Kanza, listens to their boisterous laughter and wants to join in their play. A few evenings ago, they came up from the basement dressed up like overfed robots making all kinds of weird noises. For a moment, Kanza did not know what was walking down the hall. Watching him, I could see him thinking, "What are these invaders doing inside the house?" So, he started barking.

However, once they pulled the covers from their faces, their voices turned back to normal and he heard them say, "Kanza, it is all right." His big tail started wagging again and all was okay. He pushed his wet nose into their hands, a kind of apology for his loud barking. He too loves to play--the rougher the better. But he is a dog and the others are the grandchildren. They are not all here, only four with more coming and going. That makes five of the total number of eleven.

What is interesting is that Oma and Baba's house is kind of a gathering place for now. Later they will go their own way and visit their other grandparents. Then they'll go their separate ways back home. So, what makes it fun now is that those nephews and nieces from the different corners of the country are back in Kansas. There is Jacob from Saratoga Springs, New York. He was the scout. There is Miriah from Raleigh, North Carolina, who hops from one set of grandparents to the other, both living in Manhattan. And then there are Joni and Arianna. Joseph comes this weekend. He is presently at a Boy Scout camp.

It is interesting to watch the grandkids of different ages, ranging from fourteen to four, play and interact. Anneke, my wife, is more actively involved with the driving part, hauling them to and from the swimming pool and shares in their games, making costumes, etc.

They like to come with me in the evening when I go to the farm to check on the horses and cattle. Just the other evening, Joni and Jacob came along. It was getting dark when we got to the farm so we did not stay long. However, we did stay long enough to enjoy the coolness of the evening and the warm sweet scent of freshly cut hay, prairie hay.

As they were jumping over the windrows, they suddenly remembered that they had built a stable that morning using a cardboard box. It had box stalls for all the toy horses they could find in the toy box. It even had saddle racks for the tiny saddles which came with each horse. But now, jumping the rows of freshly cut hay, they decided they needed hay for the horses. They picked up an armful of hay, which I tied together and laid in the back of the truck. They planned to cut it all up and make miniature size hay bales which they then could neatly stack in their barn. I could see it all, the downstairs basement filled with scattered hay--how much fun it is to be at Oma's!

By the time we left the farm, it was dark and a waxing moon rode the evening sky. The kids asked if they could please--please--please ride in the back of the pickup truck, at least as far as to the hard-surface road. I agreed, as long as they would sit down. I drove very slowly and through the open window, I could hear their chattering--which was like two waterfalls.

Being dusk, wildlife activity was all around. We saw a raccoon family walking the road. The little ones were hanging over the mama raccoon. A possum slinked away with his tail dragging, and two bucks suddenly jumped in front of us, their antlers still velvety. They moved along in front of the truck, clearly visible in the lights. After a short distance, they jumped back into the shrubs.

At the hard-surface road, they climbed in the cab and sang songs--crazy songs--not a bad day! I was not surprised that we couldn't find any of the kids inside the house when it was bedtime. We found them on the front lawn. They had taken the double sleeping bag and rolled it out on the front lawn and the four had crawled in to look at the moon which was standing over them in a perfect night sky.

When Anneke found them, four heads sticking out above the sleeping bag, one of the granddaughters, Joni the ten-year-old, with inflection in her voice asked if the bar was still open and if so, could they order their drinks. "Yes, four orange drinks coming up," Anneke replied. You should have heard them laugh. I even think the moon was laughing!

They enjoyed being at our place. Yes, to be with grandparents. But what they really enjoyed is that there was room to play outdoors. They built forts and hideaway places. The whole yard was their playground. I know that for weeks I will find signs of their activity. They played, used their imagination, and enjoyed what they built. They worked together at whatever they dreamt up as the next thing to do. With all the trees and shrubs, our yard is a children's paradise and that is what we wanted it to be.

July 2, 2008

The Dream

When the tornado hit Manhattan on June 11, 2008, our house was out of the tornado's path by more than a mile. Only the heavy rain caused a problem by partly flooding the basement. However, my daughter and grandson took care of it as we were away on vacation.

The basement leakage has happened before. Another time when we were gone, the basement carpets got soaking wet. My daughter had reached this stage with a kind of ultimatum, "Don't go away or fix the problem." But how do you fix a problem of a leaking basement when the house you live in stands high and dry on top or near the top of a hill? From where we live, we look down on the town.

I do know this: when the house was built, they wanted to dig the basement deeper but they reached a solid rock ledge and decided to set the house with the basement on the rock ledge as going deeper would be too expensive to excavate. That was well before we moved in, after several owners, and when Marlatt was still a gravel road--years ago.

I have explained the problem this way. When it rains hard, water is soaked up into the soil above the house and as it hits the ledge, it moves along as groundwater. When it reaches the house, it takes the path of least resistance and moves into the basement through cracks in the concrete floor and walls. Remember, this is an older house--our home.

My grandson made a wise and quick decision when he saw the water in the basement, his mom "holding her finger in the dike," by using the old Shop-Vac. He took his mom and my checkbook--remember, we were away on vacation--to the hardware store and bought the biggest Shop-Vac with an attached pump plus a garden hose. Once assembled and activated, the result made me proud. It showed that Dutch water management engineering skills, dike building and reclamation, had been passed on with the genes. His mom is our daughter, even though he carries his father's name and carries those genes as well. Today, the basement floor has been waxed and the storm has passed.

However, not all storm damage can be repaired that easily. We all remember Greensburg. People gladly would have taken water in their basements if their homes were left standing. The same for Chapman and other places that were damaged or blown away.

Here in Manhattan, the home of a friend of ours was damaged when sliding doors blew in and windows broke. They were in the tornado's path. More houses in their neighborhood were damaged, but none destroyed. And the fact is that houses can be repaired. Roofs can be rebuilt. Windows and broken furniture can be replaced. Everything except family mementos damaged beyond repair. Only memories can hold them.

What struck me after we returned and I drove up to my friend's home was the outside devastation. Their home used to be surrounded by large trees, oaks that dipped down into a steep wooded draw. There was a pleasing, winding trail through the woods to the creek. The house looked out over and into the woods.

Years ago, they built a deck and gazebo down the hill close to the house. When we sat on the deck or inside the screened-in gazebo, you had no idea or feeling that you were in town. It actually felt more like Colorado. And you could barely see the house from where the goodies were carried down--cool drinks, tea. On one side, the gazebo looked out over a small prairie meadow where deer would pass. It was an idyllic place where carefully selected plants thrived and had found their niche.

Sitting in the sunroom, you could observe a variety of birds. I especially liked the woodpeckers as they moved over and along the branches which reached out to the house. Many a happy family gathering took place on the deck, built among the cedars. You would never believe you were in town. The deep, tree-lined gorge gave it privacy. Their house was at the dead-end of a lane with several acres surrounding them.

But when I drove up to it, all had changed. The tornado had moved up the wooded gorge and destroyed the trees, ripping them out and apart. Only a few broken stumps remained. It was now an empty space. Awful devastation. It looked like a war zone, the battle of the winds and twister. Looking out over the devastation, I had a thought. The sense of place created in a beautiful, natural setting for more than 40 years through the gardening touch was gone in a mere few seconds--gone.

But in the long view, the history of those nearly four acres as part of the Flint Hills have known other storms, roaring fires, gushing erosion by the stream which runs through the bottom, and maybe other tornadoes. I am thinking back thousands of years, well before the oaks were big. Our footprints are only recent, a very short time. The 40 years enjoyed were a gift to be remembered and treasured. The land will heal. There are acorns in the ground. All it takes is patience--human patience, loving care, and the dream. However, to see aged trees once more will take another generation, at least one if not two or more.

Still, you must hang on to the dream. The land will heal. It always does when encouraged to do so.

August 6, 2008

The House

This past weekend, I read an interesting book. Perhaps I should say a chronicle. It has 306 pages with many fascinating photographs, maps, and sketches of earlier times and the present situation. The author went to archives and pushed back the history of the house where they raised a large family, well before the first family bought the parcel of land and built the original house. It was and is a well-known house, uniquely situated and loved by the family who own it now and the children who grew up in it. It is not for nothing that the author titled the book *Het Huis Waar Ik van Hou,* which translates into *The House I Love*.

The original drawings of the house are from June 8, 1922. The builder who built it seems to have said that it was the biggest construction project he had ever undertaken at that time. I have visited the house twice at different times when I was in the Netherlands. It is the home where a close friend of mine who is a faculty member here at the university, grew up. Notice that I spoke about a house first and now call it a home. It was and still is home to my friend who married an American. I dare say they both call it home and rightly so.

There are many houses built and many houses torn down without much feeling. They are just structures, a roof over our heads. Yes, this house too is a structure. It was designed by an architect. But that still does not make it special. I have seen houses built, designed by architects which make you wonder, "How could you?"

It is true, Frank Lloyd Wright was a great designer. I have heard it said that some of the houses he designed were hard to live in. But we are still talking about a house, a structure. To make a house a home, the structure does not have to be designed by architects.

The author of the book digs down beyond the basement and describes the geology--the effect of the ice age on the soil on which the house stands. It is an old Dutch landscape where my friend, as a young man, once found a part of a stone ax in a hole the neighbor was digging. It spoke of earlier people who lived there some 4400 years ago. Close by, a small, old river runs--the Drentsche Aa. It is an important part of the old natural and cultural-historic landscape. The name of the house is Huis ter Aa, or House at Aa, signifying the connection.

But for a house to have such a feeling--so much feeling that one loves it--is special. The house was home to a large family, nine siblings, parents, dogs, sheep, and geese. There was plenty of space around the house to play, get into mischief, work the grounds, weed, and pick apples in the orchard.

The book describes the rural community of Glimmen, where the family grew up. Reading it, I could see it all as I too grew up in a somewhat similar setting. Part of my youth I grew up in the Manse, the parsonage. It also had the character with a sense of place. It was my mother who turned the house into a home. Just like the author has with the house she loves, where she had lived for 50 years or more.

When I had stopped reading for a moment, I went outside and walked around our house here in Manhattan and I called my friend who lives just a few miles away near the lake. I told him that as I walked, I felt a tremendous sense of place in the garden I had created during the 34 years we have lived here. Trees have grown up so that the yard looks mature. Paths among the trees lead to open lawn and give a view to the border. In the back, I see the evening light fall on the wooden gate which opens up to a walk on the open prairie to the hilltop. There are stacks of firewood, a pole shed, nooks and crannies, and the stable. There is so much we have planted and have done to create the special feeling--including the winding driveway uphill, but above all the privacy and yet still closeness to our neighbor.

I told my friend the book his mother had written was a great gift. I know he treasures it. Whenever they return to the Netherlands, he goes home. He will step into the house and it is home as it is for the siblings and their children, the grandchildren. It is only when you live in one place for a long time that these intense feelings can grow and mature. They must mature, as over time it changes. The author is now well in her 80s.

It has been said that we in the U.S.A. move every five years to another job, another house. Moving is our spirit. However, there are exceptions, and farms and ranches belong to the exceptions. It is the farms and ranches which make homes and where the natural and cultural history can be read, to be passed on to children--grandchildren--great-grandchildren.

For instance, I know our house is built on rocks, rocks belonging to the Flint Hills. I know that Native Americans have stood on this hill looking down into the wide valley where the buffalo grazed in large herds. I know my wife, Anneke, has turned this house into a home. There are flowers. She is there.

October 25, 2000

Sleeping on the Prairie

I am trying to think back to which evening it was that I decided to sleep out on the prairie. It was Thursday night because Friday I had dinner with friends. I remember now that it was a gorgeous day with a warm evening expected, perfect to sleep outside. Wanting to do this for a long time and being alone for two weeks, I could make my own plans--not that my wife would have stopped me, or tried to, from such an overnight venture. Most likely she would have said, "I am coming too," or she might have looked at me and said, "We have a warm bed at home, why sleep in the grass?"

But I wanted to sleep in the grass. I wanted the day to slip away from me and to be slowly engulfed by the night. I wanted to watch the stars come out and later experience the light from the fading moon. All of that I wanted, and Thursday night was going to be the perfect night.

Around four o'clock in the afternoon, I threw some food and drinks, including one Heineken--a Dutch beer, in my old backpack. The sleeping bag is not much anymore after 30 years of service. I threw it all in the pickup truck and called the dog. He already knew we were going to the farm and he was ready. One mighty leap and he was in the pickup bed.

I have a habit of tying him down as at times he has jumped out when he sees other dogs which challenge him as we drive by. Also, whenever we pass a bicyclist I have to swerve away because he likes to put his head out and bark at them. The barking is not menacing, but simply for the joy of it. Yet sometimes Kanza has startled the biker as they don't expect a loud "woof-woof" so close to their head. Remember, he reaches over from the pickup bed.

Once on the farm, I threw my backpack on my back and quickly walked to the far end of the farm. It was a short mile's walk through the woods, across the newly planted wheat, and across a dry creek with some standing water in pools which the dog tried out for a bath. I crossed a brome field and walked up the hill where I planned to spend the night.

After putting up the small tent just in case it rained, I then sat down on a horse blanket and quietly and contentedly looked out over the long valley in front of me, part of which was our farm. As evening approached, the sun got lower on the horizon and without any clouds, I knew I would see a perfect sunset.

Either the dog or I must have made enough noise because the horses had come up to the fence which separated us. They are both white so in the fading evening light, I could still see them clearly and watched them stand and graze. Later, they followed a trail back over the hills to the big pond. By that time, the sun had dropped behind the horizon and I was scanning the edge of the fields below for deer. I know they are on the farm. Even when I no longer could distinguish much with the naked eye, my binoculars kept pulling in light so I could still see.

However, I did not see any deer. I saw a bird of prey swoop down but only saw it because it came into the field of vision of my binoculars. When I looked over the binoculars, I could not see a thing. It already was too dark below me. High in the sky, an airplane flew west and the last sun rays shone on its metallic fuselage, clearly visible with my binoculars, the pulsing lights and white vapor trails marking its path.

Then coyotes started to bark and yip. It did not last long and Kansa, sitting next to me, perked his ears and attentively looked in the direction from which the yipping came. Soon other coyotes beyond the hills started to join the chorus and then, all of a sudden, it was silent, and Kansa lay down again with a deep, deep sigh.

By now it was getting dark and I slowly munched my sandwich followed by an apple. It felt perfect to just be there, listening, feeling the warm, gentle breeze pick up for a moment and then die down again.

It was evening and the stars had come out. There was the Big Dipper and the North Star. I lay on my back and looked intensely into the black night. I tried to look past the stars, deeper into the blackness. I tried to understand what I was seeing. On my bookshelves stands a book I bought years ago. The title is *Music of the Spheres* by Guy Murchie and the title of that book came to mind as I was lying in the grass looking up into the starry night. I thought of some of the people whom I knew who also had looked at the stars--my brother and sister, both long dead.

As I lay there looking up, I was amazed at how many planes were passing overhead with their blinking lights. They flew from east to west, but also west to east. They flew north to south and vice versa. It was busy in the sky. And I could believe what I recently had heard--that several hundred planes cross the Atlantic every night going and coming from Europe.

I don't know what time it was when I dozed off. I really did not care. The time was mine. Fifteen miles away the sky was lit up by the lights of Manhattan but looking northeast it was dark. I rolled over on my side and dozed off. The thin sleeping bag was not much of a cushion, but the grass was.

I briefly woke up later in the night when the half-moon came up and the dog barked. Some animal must have moved below us. He soon curled up again, as did I, and we both slept until the morning light shone through the standing grass.

I put my boots back on and took a long hike across the prairie as the sun climbed up on the hills. The prairie was dry, but when I crossed the brome and wheat fields, my boots got wet from the dew. I saw a doe with her fawn. She quietly walked away from me as I watched them go and disappear over a ridge. Back on my hilltop, I munched an apple and packed up. It was nine o'clock, time to go home.

Back home I told the grandkids, "Some night you should go to the farm and sleep on the hill--you will like it."

It was my granddaughter, Joni, who said, "That's cool."

October 18, 2000

The Lake

It rained during the night last Friday. I remember rain was predicted and I was happy hearing the rushing sound at some time during the early night. The rain, a second rain on the newly planted wheat, was just what was needed. When I woke up at daybreak Saturday morning, the fragrance of still wet earth and leaves now falling to the ground came pouring in through the wide-open windows. I quickly got up and dressed.

Once outside, I put the dog on the leash and started on a brisk walk to Top of the World, only a few minutes from where I live. Walking down the driveway, I sensed the early morning fog hanging between the trees. But only when I reached the park, Top of the World, could I fully appreciate the extent of this early morning fog.

Walking along the top of the ridge and looking east I could only see fog. The whole valley of the Big Blue River and the surrounding landscape was hidden below the gray mass of fog. It looked as if I was standing on a mountain top. It was Top of the World and way below me the land was gone. It actually looked as if I was standing on a shoreline of a huge lake--a lake with no ripples. Looking east, a lazy, orange-yellow-violet sun was trying to push through some clustered clouds. The sun never really came through. But it did much later and then the fog started to lift and my big lake ran empty.

I looked for a long time out over the imaginary lake and it dawned on me that this is what Kansas must have looked like millions of years ago or thousands of years, if and when there was a big flood.

Only a few hilltops rose above the silent fog, making it look as if there were small islands drifting in this sea. That early in the morning there was no traffic and the absolute silence made this watery scene, an inland sea, seem even more real.

I honestly would not have been surprised if, out of nowhere, a rowboat would have come with the rhythmic knocking sound of wooden oars on a wooden boat quietly pulling through the water. A boat with a man, a fisherman, and his dog, slowly rowing to shore. Standing there, I would have helped him pull the rowboat on land and tie it to a rock. The man could not have spoken to me; it would have been a ghost. And he did not speak to me. He never came. But it looked as if he could. That is how much the whole valley had changed into a lake of fog.

As the sun slowly pierced through the hazy clouds, the fog started to lift and more islands appeared. Clusters of trees slowly rose above the fog as did lower hilltops. By the time I walked off the ridge, you once more could look into the valley, and Saturday morning traffic started to pick up, breaking the spell. I only wished I had brought my camera to record this unbelievable scene--this huge lake or inland sea.

This view is recorded in my mind. I wonder if I will ever see it again. The time, weather conditions favorable for a thick fog, and being there all will have to fall in place, including the early morning sun just trying to break through the clouds. But even if I never see it again, I will remember it from now on. And every time I walk the ridge on Top of the World and look toward the east, in my mind I will see a large expanse of smooth water covering the landscape, covering the land, and I will be listening for a rowboat coming to shore!

Seasons

Where is East?

January 6, 2010

The Lone Chair

Looking back to the days before Christmas there is always the feeling of a rush. There is so much to do and so little time in which to do it. There is the anticipation--Christmas Eve, Christmas morning, and Christmas day.

I refuse to get caught up in the rush, which is basically a rush to shop under the motto "The more you buy, the more you save." Because I did not grow up with it when I was a child or teenager, that aspect holds little meaning for me.

Following Christmas are the days between Christmas and New Year. Those six days hold something reflective for me. I really like them and seem to live them with my inner soul. There is something melancholic, but not sad about them. Those are the last days of the year, as if placed between apostrophes. No accolades. They are quiet days, days to be lived like you taste a good wine. This year, the weather helped set the stage with the overcast gray sky, cold, and snow.

The day before the New Year, I took a homemade loaf of bread to a friend who lives in the country on a farm. It was snowing when I drove down the icy driveway. The freshly falling snow did not stick to the road but drifted and floated in front of me as I drove along. I loved it. Tomorrow would be New Year's Day. This was still 2009. The very special last days. When I turned off onto the county road, it was covered with compacted snow and my truck moved along silently. I was in no hurry and did not want to slide off the slick road.

Angus cattle stood in the fields--black on white. They were looking among the remaining cornstalks or were standing around the rolled-out bales of hay. A couple of horses stood with their tails to the wind, carrying snow on their backs. They had shelter but seemed to prefer the openness with falling snow.

I passed a harvested cornfield with the stalks marking the long rows curving along the creek; good bottomland now white. A flock of crows flew off. They did not hurry. Black on white with a lazy wing beat. Their flight was erratic, yet they stayed together as if being together kept them warmer.

It was a beautiful drive with the trees darkened and snow clinging to the top of branches. I drove past our farm and thought of Robert Frost's poem, "Stopping by Woods on a Snowy Evening." There is this line:

"Whose woods these are I think I know.
His house is in the village though;
He will not see me stopping here
To watch his woods fill up with snow."

He then talks about the horse which does not understand why he needs to stop. The sleigh bells jingle and he travels on with miles to go. I did not stop, but I could see that no one had entered the farm through the gate. The snow was deep and clean.

When I got to my friend's house, I parked the truck so as to not get stuck as the snow was piled up high. Paths had been cleared to the doors. I knocked and called. No one answered. I stepped inside and opened the kitchen door, calling once more, letting him know I was there. The collie stood up from its blanket and greeted me. When no one stirred, I placed the wrapped bread on the table and closed the door. Outside, healthy cats and cute kittens skittered across the snow.

I walked back to my truck again, passing the single garden chair standing in deep snow. It was a metal chair with snow on the seat. Today no one would sit in it, watching neighbors go by. There used to be two chairs placed against the porch and sometimes three if friends or neighbors stopped by. Only one chair now. The old farmer, the father, died last fall. He was well into his 90s. No complaints there and he died as an old farmer should, working--driving his tractor, mowing tall weeds.

When I stepped into the warm cab of my truck and backed out, I looked once more at the lonely chair standing in the deep snow--one chair. My thoughts lingered on the old farmer. He was a great man, though small in stature. I feel privileged to have known him. I did not feel sad. He was well in his 90s, what more can one ask for? For me, the last day of the year, the old year, is now going by fast. You could count it off in hours. That midnight, in seconds--10 - 9 - 8 - 7 - 6 - 5...

The new year brought more snow and a call from my granddaughter's husband that he and his buddies wanted to go and cut more firewood on the farm. With the low temperatures and the ground frozen, I had no problem with it. Generally, I don't want to drive on the land.

On Saturday, I drove out to see how they were doing. It was now January 2, 2010. The new year had started and the day had a different feeling. I parked at the gate and followed the deep tracks through the fresh snow. They were cutting the broken-up huge oak which had blocked a trail. Walking up along the woods, I smelled smoke and stepping out I saw fire. The young guys were sitting and an older man was standing around the fire in the snow, all eating hot dogs.

The whole setting reminded me of Robin Hood with his merry men celebrating under the mighty oaks in the king's forest.

Where the lone chair was the Old Year, here the banter, laughter, and working were the New Year!

February 19, 1986

Kansas Mud

We are now getting into the season of mud--sticky mud--and gumbo.

You are missing something when you live only on concrete. When you walk out of your house onto a concrete walk to a concrete driveway, drive on a concrete road to your work, then walk again on a concrete walk to enter your place of work, you are missing something. You are missing something if you never are able to put your shoes in the mud--Kansas mud. In all reality, only a very few of us really live on concrete or any other hard surface and most of us probably do encounter mud somewhere.

It is at this time of year that good old, sticky mud seems to be around everywhere, especially if you live and work in the country. It happens when the hard surface roads of winter first freeze then start to thaw out, especially if some snow comes sloshing along to make it really good. A warm day thaws out the surface and you have it--fresh Kansas mud.

Wherever there are gates and cattle or other livestock passing through, it suddenly becomes a mud bath in which any hog would rejoice. What was a hard surface a week ago when temperatures dropped well below freezing is now oozing with spring fever.

The farmyard where tractors and trucks turn around, back up, and go in circles is slowly turning into a mud mire. Oh, there is no question that the 4-wheel drive vehicles can get through. You know that even with mud up to their axles, any respectable 4-wheeler should make it.

This is the time of year that if you want to keep mud off your boots, you drive from the kitchen door to the barn because you might lose a loose-fitting boot and suddenly find yourself standing barefooted in the cold mud. Of course, good design of the farmyard with adequate drainage and good landscape design around the home would eliminate much mud.

Now is when you fight the battle of the kitchen floor. You can't keep it clean because if the man doesn't track in the slippery, sticky stuff, then the kids or dogs do. With the good weather, it is hard to keep the little ones in so you send them out. As you watch them go through the yard, you see them climb onto their tiny, toy John Deere tractor and where do they go? They steer right into the mud and soon you see them pushing, pulling, and trying to get their stuck tractor out. The power only comes from their little legs and trying as they do, the tractor soon sinks down to its axles. As you watch them, you can see the little farmers stand around the object of their concern with great satisfaction and happy smiles. They managed to get it stuck in the mud just like daddy.

Now comes the best part, trying to get it unstuck. They look for a piece of rope and somehow tie it to the front of the tricycle and someone is chosen to pull. It often is the weaker one who has to pull because the bigger brother has to drive the tractor.

By now you have watched so long that the dishwater has turned cold and you decide to finish cleaning up the breakfast table. You don't need to look any longer because you know what will happen. The rope will break or slip loose and whoever was pulling will slip and land in the soft mud--Kansas mud.

You better get ready for the muddy season. It is an important part of the year. It heralds spring. Even though more cold snaps may turn the soft mud into a landscape of miniature craters, you know that before long, all this mud will once more be dry dust when the summer heat is upon us.

Now you better throw some old feed bags out near the steps because I can see the big and the small man coming in for coffee and they have mud on their boots--clean Kansas mud!

February 24, 1999

Isolation

Monday morning brought a light snowfall to Manhattan. It was nothing big, just some wet snow which blanketed the lawn white, but that is all it did. The rest of the day was dull gray and as the snow melted, it all became slush. Of course, slush on asphalt or concrete is not too bad but slush in the country means mud--black mud, red mud, sticky mud, slipping mud.

Before the 4-wheel drive trucks, mud could isolate anyone living in the country along a dirt road. Today, the 4-wheel drive vehicles with good mud grip tires will claw their way through the deep mud tracks I have seen along some county roads--roads I don't care to enter until frozen over and even then, you have to be careful. But once dried out and after a few passes by the road grader, the isolation is broken and one can drive back into town. It does not take too much to realize that before the 4-wheel drive vehicles, isolation was a common problem in the winter if you lived in the country. It is very interesting to hear people's reactions to that perceived isolation.

The Marianna Kistler Beach Museum of Art here on campus presently has several paintings on display, among which stands out *Sun Dogs* painted by John Steuart Curry in 1930. It is oil on canvas. I have seen it several times during brief visits to the museum but went back the other day to take a better look at it. It is a canvas roughly two feet high by three feet wide. The scene is winter in the country. It shows a small red barn and a gray house surrounded by snow--swirling snow--and a snow-gray sky with the dancing sundogs. The landscape shows a dark ridgeway off in the distance. In the foreground is a farm road showing some bare windblown soil leading to the distant red barn in the valley. Besides the brown soil, there are some brown, broken grasses in the foreground, but otherwise, it is all snow--snow and isolation. Snow drifts closed the road as there are no tracks.

My wife, who is a docent at the Beach Museum, had an interesting conversation with a group of visitors whom she was showing around. When they stopped in front of this painting, several of the younger women in the group remarked about the terrible isolation of the homestead with the red barn in the snowy landscape. You could watch them shudder at the thought of living so isolated and being further isolated by inclement weather, after a winter snowstorm.

As the young women talked and thought more about the isolation, some of the older women shared their feelings about the picture. They said, "You look at it wrong. Being isolated by a snowstorm like this brought family togetherness which was enjoyed, and in a quiet way, even celebrated."

There was no way to leave the farm. Yes, chores had to be done but everybody stayed home and it could be very peaceful. The older people remembered the very positive feelings associated with a snowstorm and the following isolation. They remembered the forced togetherness which was enjoyed.

I remember a Thanksgiving in Virginia many, many years ago. My brother-in-law who lived in town came to visit us with his young family. We lived only a short distance out of town, but in the country. During the visit, it started to snow and it snowed hard. When it came time for them to go home in the evening, we realized that the unplowed road might be a problem. Also, some snow had drifted across the long driveway which made us decide, "Why don't you all stay?" And that is what they did.

We had a great extended visit with the kids playing in the deep and drifting snow. There was no way to get out. There was no need to go anywhere and no one wanted to go anywhere. That was exactly what the older women were talking about. They smiled when they listened to the young women talk. The older women knew what the young women were missing--quiet time, family time--forced on you by isolation during a snowstorm. Lonely, but not alone!

July 1, 2009

Flash

"Paps, you rode that horse in your mind all the time!"

The god of thunder--called Donar in the mythology of the old country, called Thor in Norway and Scandinavia--wanted him, needed him, and took him. He took him by throwing one lightning bolt down to the earth. He aimed well.

Coming out of the woods, I entered the range by the lower gate close to the big pond and briefly I smelled the scent of death. I wasn't surprised as deer and possums live there and animals hunt and take down their prey.

I walked to the pond, along the draw, feeling the coolness of the trees growing there. Close by I saw the horses, the bay and a white. I only saw two but expected the third horse to be hidden by the dogwoods. I walked around the pond, crossed the running rivulet, and stepped out onto the open range. The horses had moved up the slope, over the ridge, and out of sight. I followed them, noticing the good grass underfoot. The cows slowly circled back to the pond as I walked onwards. I wanted to see all three horses. They always graze together--Poptart, Flash, and Misty.

When I got to the top of the hill where the old buffalo wallows hold water, I saw them standing in the distance. But I still only saw two horses. I whistled and they raised their heads. They both came my way for their handful of grain and some conversation. Misty walked easily; her foot having healed. As a matter of fact, I had watched all three of them canter and then race up the hills a few days ago. Once on top, the horses stopped, turned their heads, and snorted, showing off their beautiful form as silhouettes. I love them. Each one of them is a character, and they know it.

While I gave them their grain and felt their soft lips gently eat from my hands I talked, asking, "Where is Flash?" I whistled. Flash always grazes next to Poptart. They are mates. Poptart has carried several of his foals, beautiful foals. We had two. Thunder was born during a fierce thunderstorm, hence his name. And a year later, Misty was born during one of the foggiest mornings ever over Manhattan. Thunder was a boy, a tremendous jumper. He managed to jump out of the coral as a little foal leaving his mother behind. I have often wondered what made him do it--the independence of his father, Flash? Misty on the other hand is laid back and easy going.

I started walking across the hills in larger and larger circles, looking into the valleys, whistling. At one time I thought I heard a snort, typical for him. But it was the cows. I walked for a long time. Then I went back to the pond and once again scented the smell of death. This time I took it seriously. It led me up the hill once more to the top.

I followed the smell until I saw a raw hump in front of me. I walked up to it. There, lying along the ridge against some sumac was the carcass of Flash. His legs were stretched out. The vultures lazily drifting high overhead; the coyotes, flies, bees, and maggots had been busy. His rib cage was torn open and I was tempted to look for the missing rib--Flash was an Arabian, at least three quarters. He had the typical short rump and the soft, intelligent, gentle eyes. Those eyes were now ripped out of their sockets. His beautiful mane and tail were torn off and lay matted in the scorched grass.

Flash was no longer there, only his bones, skeleton, and stench. I looked at what remained of his face. It reminded me of the horses of the Apocalypse. But no, that is not where this horse is. Flash lived the full life of a horse. He knew freedom. Whenever he was turned loose on the range, he would kick up his heels and storm up the hills, stop on top, turn around, and look at me standing below him at the gate.

It was my son, Martien, who said, "Paps, that horse was made for you. His temperament was just what you needed to be challenged." And he was, but Flash also knew how to give. A few times he would lay his head over my shoulder. Those were special moments. His lips would take the salt from my hands. Other times we would read each other's eyes, his big, strong yet soft eyes.

As I said, the god Donar wanted him. Flash must have been grazing on top of the ridge as the fierce storm rolled in with dark clouds and wind pushing the rain. His tail must have blown with the wind. He may have raised his head, taking the terrible blow as he stood there followed by instant death. He did not crumple, he fell over. He was Flash, killed by a flash of lightning.

He will now live with me forever. Whenever there is a thunderstorm roaring across the hills, I will hear him galloping in front of the chariot, thundering. I will see him in the rolling black clouds with lightning flashes.

He is running with the storm. There he is alive and I will ride him--in my mind.

October 10, 1985

Autumn--The Season for Fall Storms

It is autumn, the time of year for storms, rain, and wind. The time for leaves to change color and then be blown off, either slowly drifting to the ground on a nice fall day or pelted down by fat raindrops and slapped on the ground or against a building or car.

It was a stormy autumn day, the Sunday morning a week and a half ago, that my then eight-year-old son said, "Look at those branches arguing and fighting together." We were sitting at the breakfast table and it was cozy inside. Outside there was a roaring wind and trees were bending and shaking. The one particular tree he was looking at I planted ten years ago. A seedling locust coming up as a volunteer; it has no thorns and no pods. It has grown from a mere whip into an attractive tree at least 25 feet tall.

I planted the tree for many reasons. To provide shade, a lacy shade for that part of the house, to provide branches for birds to perch and to find their place in the pecking order near the feeders, to look at a tree from close up, and to watch it grow over the years and become part of our landscape.

My son grew up with the tree and from where he sits at the breakfast table, he must have looked at it every day. But seeing it and looking at it are two different things. I now know that he has seen it. He has observed it and saw something which was funny to him.

While eating his breakfast, he observed two low hanging branches being battered by the wind. They were sweeping and bending away and toward each other. They were flicking each other with their twigs and then torn apart again to bend away from each other as the strength of the autumn storm increased or abated. To Martien, the branches looked agitated and that is when he said, "Look, those branches are arguing together. They are having a fight."

Indeed, the way the branches were thrashing about it looked like two boys on a school playground getting into a tussle with arms and legs flailing about. Such a boys' fight holds a lot of laughs as legs and arms move like windmills and there is much puffing but very little damage done. It is no more than good exercise and when it is all over, the stalwarts are buddies again and playing as best of friends.

And that is what the two branches looked like to him as he was watching the storm blow the tree about. I watched the gleam in his eyes and I could see him thinking and nearly say, "Go for it, get him," as boys will do.

It is not that I like to see boys fight. Oh, a good clean fight is all right. As I said it is good exercise and seldom is any harm done beyond a torn shirt. However, what I did like very much was his imagination. The ability to see the two sweeping and thrashing branches as if they were arguing, being agitated beings. That I liked because it proves to me that he can be creative and not only that, he can express it and say what he sees. Of all the kids, he may be the one most like me, this caboose.

Of course, one can teach children to be imaginative, but some will be more so than others. The environment in which one grows up may lend itself, more or less, to it. However, I don't think one can say that one environment is better than the other, as long as the stimulation and initial interest is there. In our case, I have planted trees, shrubs, and created a landscape for adults to enjoy and for kids to play in. And we have watched them play and grow up and enjoy this sense of place.

November 7, 2001

A Fall Day

I don't know how many more weekends of gorgeous fall weather we will have because one of these days, winter must start. However, for now I am thoroughly enjoying the pleasant weather.

This past weekend, with temperatures well into the 70s with no wind and clear blue skies, showed Kansas weather at its best. Even a flat tire on my tractor did not upset me too much except I had to take it off and take it into town to have it fixed. When I was asked in how big of a hurry I was, I cheerfully said, "No hurry at all, as long as the weather stays good."

The next morning, I picked the tire up and returned to my job on the farm mowing field edges and clearing trails. I mowed the brush, rough-leaved dogwood, and wild plum with some sumac thrown in for good measure along the edges of the fields to make sure the shrubby growth did not invade the brome meadow. I was not surprised to have picked up a thorn earlier because between the shrubs and thorny locust sprouts there are plenty of sharp objects to puncture a tractor tire.

Mowing along the edges, I scared a few rabbits. However, they were safe as the dog was following other scents--deer! He chased a doe over the hills without a chance of ever catching up with the graceful, loping deer. A little later a beautiful buck jumped out when I got too close to where he was bedded down or had stood courting the doe. He too bounced across the field, up and over the hills with his white flag waving. He was not in too great of a hurry either, apparently knowing that tractors do no harm. He looked like a nice corn-fed buck, solid.

My neighbor's cornfield was harvested this weekend. Another sign that somewhere winter is waiting to come down. The cornfield is just across the fence. From my tractor seat, I could see the blue farm truck filled to the brim with yellow corn while the combine was going around and around, grabbing the corn cobs and husking them, releasing the corn into the bin on top. Together it made for a perfect fall day--harvesting corn, maybe harvesting a buck, tidying up the edges of farm fields.

After a little while, the dog came back from over the hills--panting. He no longer had the energy to follow me around the field but plopped down in a shady spot and watched me from there. When I moved away from the back of the farm, he found his way to the pickup truck and lay down, knowing darn well that sooner or later I would end up there.

I sometimes wonder how smart the dog is. He knows the boundaries of the farm. Although earlier that afternoon, he sauntered over to the neighbors when he noticed them standing near the loaded truck. They know the dog and he just paid them a visit as I waved. I hope he did not lift his leg against the truck tires. That irritates the dickens out of me. With all the trees and fence posts, why lift a leg--a dog leg that is--against truck or tractor tires? All dogs do it and that is why they keep doing it, marking the spot, "I live here." The trouble is, trucks move which must be rather confusing to the wandering dogs.

With the harvest just about completed, farms are getting ready for the colder weather to come. Working with machinery is still a pleasure. From my tractor seat, I could see the autumn color of the prairie over the hills. Along the creeks, the cottonwoods had lost a lot of leaves by now, but in the tops there was still a yellow crown standing out against the clear blue sky. The puddles of standing water in the creeks had leaves floating in them, mostly browns, but a few reds and yellows. With no wind, it was very quiet, except for the steady sound of the tractor pulling me along.

A few days ago I had noticed a large flock of crows, at least 50 of them loafing, hanging around. They made a lot of ruckus, doing nothing except crowing that I was there. I was glad I did not see them over the weekend. As winter closes in they will be back, hanging close to the ground, letting the wind, the air current, sweep them up or just hang for a little while like a stalling hang glider.

For me, they are a winter bird. I like to see them against the snow--black on white. But for now, I'm enjoying the day. It cannot last much longer.

November 23, 1994

Fruit of the Gods

Pumpkin pie, cranberry sauce, and all the other trimmings. I can taste it already! Pumpkins we can grow, but the cranberries have to come from somewhere else. Yes, turkeys we have too. I saw a big flock the other day, but cranberries, no we don't grow them. I wonder if perhaps we could change that tradition with some other special fruit that ripens this time of year and grows here in Kansas.

I propose we make our sauce from the Diospyros virginiana. The Greek word Diospyros means "fruit of the gods." Don't let the Latin word virginiana bother you. It simply means that the Diospyros--"fruit of the gods"--grows in Virginia. But let me tell you, it grows in Kansas too. I have seen it--single trees and large trees in the center and then all the little ones clustered around it. It is a pretty sight.

My son has a good cluster growing on his land and I plan to go down and pull some of the small trees up and take them to our land and plant them on the farm so that I too will have this "fruit of the gods." Because the trees are dioecious, meaning male and female flowers are on separate trees and only the female tree bears fruit, I better plant several trees to increase my chances of having a female tree and having fruit.

Of course, the wood--the quality of the wood--is the same for both trees. The dark heartwood of this tree is known as American mahogany. It is used where fine, close-grained wood is required for a beautiful finish.

If digging up small trees is too much work, then collecting the fruit will give me seeds. I can use the fruit pulp to make the sauce and splash it over my turkey. But the seed planted readily germinates and in time will grow into an attractive, strong tree. The tree is long-lived and in a good location will grow into a specimen tree reaching an age of 100 years or so.

In case you wonder what tree this is which bears the "fruit of the gods," let me tell you what Captain John Smith, one of the early settlers of Jamestown, Virginia, said after he tasted this fruit. "If it is not ripe, it will draw a man's mouth awry with much torment."

Now, come on Captain Smith--be a man. A little puckering does not hurt your mouth, but then again, you better wait and make sure that this delicious fruit is ready to be eaten. Because if you taste it too early, your mouth will pucker up

You have to know this fruit. When first green, you don't touch it; not when it turns amber or livid orange. No, you have to wait until the skin of this delicious fruit is all wrinkled and the fruit actually looks unappetizing--looks rotten. It is when the fruit is mushy that it becomes the "fruit of the gods." When ripe, the fruit has a wild, sweet taste. I am, of course, talking about the persimmon, the wild persimmon which can be found growing in eastern Kansas.

A few weeks ago, when passing by a cluster of persimmon trees on campus, I stopped and enjoyed some of the fruit. At the ripe stage, they are messy eating, but they were good. If you know of a fruit-bearing persimmon tree and you want to try a persimmon sauce for Thanksgiving, you have to hurry up. Possum and other wildlife love the sweet fruit too, so there may not be much left. But what is left is surely ripe by now, so don't hesitate. Knock off the last fruit or pick it up from the ground underneath the tree. If it looks bad, it will be good to eat.

The common persimmon is an interesting landscape tree. It is a native, good for naturalizing on golf courses, parks, and even home landscape but keep it away from the patio or where cars are parked. The tree prefers moist, well-drained soils but will do well in low fertility, dry soils. On older trees, the trunk has a thick bark with an interesting blocky pattern. Branches are strong and open, creating a slender oval crown, reaching 60 to 70 feet in height with a spread of 30 to 40 feet.

I called my grandson in Holton and asked him, "Kit, would you collect seeds from the persimmon tree behind the barn? I will pay you well for a full bucket of this fruit of the gods."

All he said was, "O.K." I don't think he knows how delicious the persimmon is--on the other hand, maybe he knows.

Anyway, I want to plant some seeds and a bucket full of this drooping, drooling, juicy mess should give me enough seeds to start a small orchard of the persimmon--the "fruit of the gods."

December 7, 1994

Planting Persimmons

Planting persimmons can be fun--that is if you have help and keep your spirits up!

They were spending the night with us and I told my grandson, Christopher or "Kit," to get a good night's rest because I needed his help planting the bucket of persimmon fruit he had gathered. It was more than a week ago that his dad and he brought me the bucket of persimmon fruit. However, I had no opportunity to plant the pits yet. As a matter of fact, I had left the bucket on the farm under the roof of a shelter, half expecting that some raccoon would find them and just sit down for dessert, gobbling up all the delicious ripe fruit, leaving me nothing.

Saturday morning Kit and I drove out. Oma had put some hot chocolate and cookies in a bag for us to snack on when we got hungry. His nine-year-old face had a big grin. He had me all to himself and would not have to share me with his brother or sisters, or anyone for that matter.

The morning was very pleasant. It was warm, no wind with a slight fog hanging over Manhattan. If you did not know better, you would have thought that it was spring. Ha--but winter is yet to come. I was glad and happy with the good day because it would make it easier to quickly dig a hole with the hoe and drop some of the fruits in and tamp the soil back. It is much harder to plant in frozen soil--you can't do it--at least not a good job!

On the way out, as we drove across the mile-long dam at Tuttle Lake, Kit tried to hold his breath. He did not make it all the way and gave up with a big burst of his lungs. He probably could have made it if I had not kept to the speed limit. However, the breath-holding experiment, or record-setting for the longest distance of holding your breath, was not worth the risk of getting a speeding ticket. But I was tempted, especially when we were more than halfway across!

I knew he would not blackout in the attempt because his eyes and whole face were grinning!

When we got to the farm, we parked the truck on the little meadow near the beaver pond and picked up the bucket with the persimmons. They still looked like ripe--very ripe--persimmons and no coon or possum had dipped into it. We walked past the young plantings of trees beside the creek and proceeded to plant the persimmon fruits as we moved along. I dug the holes; he planted the seeds. We planted two in a hole. Each fruit has from one to eight seeds, so we planted at least two seeds per hole or possibly eighteen. I know it is a rough way of planting seeds, but it is easy and also similar to nature's way.

It is not only a rough way, it quickly became a rather messy way because as we worked our way slowly to the bottom of the bucket, the fruits had gelled together and it became more tricky to pick out two clean fruits. Soon Kit and I were singing a song about how messy this job had become. We picked up a refrain à la Mary Poppins that this planting job was the messiest, the stickiest, the droopiest, job there was.

As our planting progressed and the bucket became lighter, the fruit became even messier and our refrain became longer with more words added describing this non-glamorous job. We walked the lowland and planted along the fringes of the woods. Nine-year-old Kit said, "I will surely have enough persimmons to pick when I grow up."

To which I added, "I plan to be around."

I liked the thought that he expected to be harvesting the fruit of the trees we were planting. Even though our way of planting was a messy way and success was not guaranteed. If only ten percent germinates and if only one percent survives, he will surely remember having planted these trees with me. He might even remember the crazy song we chanted as we moved along, planting seeds on a gorgeous Kansas day. Planting persimmons is the most ridiculous, the messiest, the silliest, the stickiest, the droopiest job there ever was!

Somehow, between the laughter and singing, the bucket emptied and we felt pretty good. Having finished the job, we decided to warm up with the hot chocolate and go home. On our way back, Kit again wanted to hold his breath when crossing the mile-long dam. This time he did not make it far at all because I made him laugh--and laugh he did.

When he could talk again, he did say "Baba, I love these foggy days." And I told him it reminded me of the Dutch weather which I liked so much. We call such a day a "'Polhoeve dag" after the old home place in the Netherlands. It had been a beautiful day to plant persimmons together with your grandson.

November 5, 2008

Ready for Winter

A Sunday afternoon in November; it is four o'clock. A perfect Kansas day to sit outside on the deck. The sun goes behind the big Bur oak which I planted years ago as a very small tree. I planned and planted right so that once grown, the tree would give afternoon shade. All I had to do was wait. Wait until the tree had grown and was tall enough to reach the path of the sun rays.

What is also interesting to me is that the form of the crown of this Bur oak grows more oval than the shape of another Bur oak I planted as an acorn. That tree, the acorn tree, has a very robust crown that reaches out to the sides and would give more shade on the deck. But it stands where I planted it next to the shed, also an important spot to me. That shed was placed many years ago, selectively where a shed should stand, out of the way yet very shed-like, if you know what I mean. It stands back in among the cedar trees slightly downhill on a slope.

A funny story with the shed is when I started digging the holes for the nine poles, I hit solid rock and had to pound with the heavy crowbar and sledgehammer to break the rock so I could place the poles. The poles were salvaged power poles my son found and gave to me. Anyway, soon after starting that project, my wife and I took Australian friends who were visiting us on a scenic tour of the state and into Colorado. I told my son, Ian, that I would like him to finish digging the holes, at least two feet deep or deeper. It would be hard work, but strong as he was I saw no problem. I marked the places where the holes needed to be.

When we returned from a great trip through western Kansas visiting places like Castle Rock and others to show the diversity of our state and Colorado, I walked to where the shed was to be and yes, nine perfect holes were dug--deep, smooth, straight. Seeing the piles of rock near each hole, I knew it had been a tough job, hard work. I was thankful that this time I did not have to fool with it but that my son took care of it. Well done! I learned how well done once I stepped inside. My son does not fool with work like his old man. On my desk lay a bill for $75.00 for renting a pneumatic drill. Ever since then, I have not worried about that son. He is an excellent engineer. He uses his head. Yes, Papa!

Anyway, we built the shed and I have enjoyed it ever since. Storing boats, a trailer, lumber, and horse feed in it. Each time I step into the shed I pass the acorn oak, now a big tree after 32 years. This is a neat corner on our property, with a green wooden gate leading to the pasture where a trail through the now tall grass takes you to the top of the hill. From there you can look to the far hills as the leaves are dropping. The two frosts we had a good week ago helped to open up the vistas.

With Sunday's temperatures in the 70s and Monday's even higher, you wonder if we really needed those damaging frosts. I still had all my potted geraniums out in the open. There were more than 50 and boy did those frosts nip them. I covered them up, but the leaves turned black and curled. I know the plants are not dead as new leaves are sprouting. I moved them all into the garage using the dolly I specifically bought for that purpose. The pots are big and filled with good soil. Lifting them gets more difficult. Having learned from my son, I bought that dolly! Yes, Papa!

I spent the next two days repairing the roof and improving my greenhouse, built of heavy plastic and frames from a huge crate. It worked last year and I expect it will work again this year. Where else can I overwinter more than 50 large pots of geraniums? It has to work. Having replaced the roof plastic and improved on the slope, water and melting snow will run off better now.

Next, I looked over each geranium pot and removed the burned leaves, added soil, and pruned them back. By the time I was done, the garage was full of cuttings--red, pink, purple, and white geraniums. The fool I am, I sorted and saved the best cuttings. The rest were swept together and hauled off to the brush pile--two wheelbarrows full. Oh, that could have made another 50 pots--big containers--easily!

Geraniums are so easy to propagate. You can stick cuttings in water, or as I do, stick them deep into a good organic soil. Remove some of the leaves to help the cutting to set and lo and behold, there will be another plant! I generally place five cuttings in a pot, but I have put up to twenty in one pot--a very big pot! These large pots of flowers are the eye-catchers in the yard, the landscape around my home in the summertime.

Once cleaned, I placed the geraniums inside the greenhouse. I watered them. It looked neat. It smelled like a greenhouse. I will set the heater up ready to go--winter can come. I am ready. The geraniums are ready. The trouble is, it is 70 plus degrees Fahrenheit.

Come on--I didn't do all that work for nothing!

November 22, 1995

The First to Speak Up

The one color photo is the last I ever saw of him.

He made a mistake--a big mistake. He should have kept his mouth shut. Of course, we recognize freedom of speech and yes, he used this freedom. And it is true many of us never use it. Some of us never speak up. But this one time he should not have spoken.

The gentleman was all gray. He looked magnificent--a few black spots among the gray which actually enhanced him, made him look even more attractive, more aristocratic, dignified. He did have nobility in his bearing.

I regret that I never met him in person. I would have liked to listen to him. I'm sure he could have told me many things about his ancestry. How they used to roam the country in large groups, staying close together for safety, paying close attention to their surroundings while they searched for food.

He admitted that he had gotten soft, that his life had become a pampered life and that may be the reason why he did speak up. I will never know.

Like I said, I never met him but I was given a photograph of him where he stands with his friend. His friend is still alive and that is simply because he kept his mouth shut. He did not speak up. It bothers me. Is there a lesson to be learned here? When asked a question, should you speak up or act as if you did not hear the question--ignore it?

But this country was not founded on silence. Maybe his friend was thinking, just thinking, about how to answer the question. Maybe he was polite, not wanting to speak first. I know I often jump into a conversation and perhaps I should learn from the friend and allow other people to speak first. But sometimes you feel so strongly about an issue that you forget.

The photograph shows both gentlemen. They are standing together in the sunlight all dressed up for the occasion, the gray gentleman and the other in fall color, brown-bronze. They are looking away from the camera, showing their stuff all fluffed up.

If only I could have talked with them. I might have learned why the gentleman in gray opened his mouth. I bet his friend does not even know and I wonder how he feels. He must be sad that his friend is gone and will never come back. They cannot share their stories ever again of their aunts, uncles, nephews, cousins, and the other family members now long gone.

Looking at the photograph of the two friends, I wonder if the surviving friend has had any regrets of not speaking up when it would have made a difference, a big difference.

What was interesting is that earlier, one of the guys did not show off his beautiful coat and actually did not look that impressive at all. That was the man in the gray coat. And on second thought, I now wonder if that is perhaps the reason why he spoke up, having been given the chance. Because when they first met him, they said he has nothing to show, just looked ordinary. There was nothing dignified, important about him. Whereas his friend strutted up and down, showing off in his suit.

I don't know, but that could be the reason why he opened his mouth. I am positive he knew what was coming but then again, maybe not. Maybe the good life had totally addled his brains and he could not put two and two together anymore. Maybe he did not understand the meaning of love. On the other hand, maybe he did. I will never know and his friend won't talk about it. He is not talking, not talking at all. Maybe he is smarter than I thought he was. Although when he bragged about his beautiful coat, it could have been his big mistake.

The last I heard, the survivor is contemplating writing a book. I don't think it will ever be published. As a matter of fact, I know it won't because he is either too afraid or just too smart to open his mouth to talk. He has learned a lesson.

It all happened a few days ago. I did not read it in the paper, but I'm sure it happened and happened over and over again. They came and opened the door and looked at the two gentlemen and asked which of them would volunteer, and the man in the gray suit gobbled--"gobble, gobble, gobble"--and that was it.

The color photo is the last I ever saw of him. When asked who volunteers, he was the first one to speak up. He was taken away. In a few more days there won't be anything left of him but the feathers in a brown bag--someone wanted them.

Maybe there is some solace in the old saying "What you eat and drink today, walks and talks tomorrow." Maybe.

Whatever you do, wherever you go, you better think before you talk or open your mouth. You should have kept your beak shut, you dumb turkey.

Happy Thanksgiving!

December 24, 2003

Christmas 2003

We all know Christmas stories and legends which are told or read to children, even adults. There are legends of the Christmas tree, as well as the Christmas celebration. The Christmas happening, the birth of Christ is of course the focus of Christmas. But like I said, there are many Christmas legends that are told over and over and become a tradition.

This past Saturday, my daughter and I drove out to the farm to select a Christmas tree. For years I have cut a cedar, of which we have plenty. However, to find the perfect tree is not always easy. There is a legend about the "not so perfect" Christmas tree. But I was looking for the perfect, or near-perfect, tree.

During the fall as I worked on the farm, I already had looked around and made some mental notes of trees I thought might do quite well. So after I parked the truck, my daughter and I set off for a walk across the farm to select the tree. The first tree I pointed out was a "no" tree. "That is not a good Christmas tree, Papa." So on we went.

As we walked, she said, "It has been quite a while since we have gone out for a Christmas tree together." And she was right. It had been quite a while. I treasured the walk together and if you remember this past Saturday was a perfect day.

Other years I have had to walk through snow or bundle up against the cold. But this year it was pure pleasure to walk the farm looking for a tree. Little snow was left where the ground was shaded. We recalled a tree I had cut years ago which still had an empty bird nest in it. We carefully took that tree home and set it up and decorated it, leaving the nest intact.

Talking, we recalled how in Virginia we would go to a friend's farm and with the sled, haul the tree down the hill and load it onto the car to drive home. That was a time when all the children were part of the Christmas tree expedition. They were all still small then.

In Virginia, it quite often had snowed and we had a white Christmas. And because I have always put the tree up rather late, the excitement of the children was therefore greater, which I liked.

I showed my daughter another tree I had looked at and thought worthy to be our Christmas tree. It stood somewhat sheltered, yet free so that the branches had grown in all directions and the tree was rather well balanced. Also, and very importantly, the needles were green and not yet red. For after all, I selected a red cedar. When I stroked the branches, they felt soft, not dry and sticky. My daughter took a good look at the tree and approved. "That is a nice tree, Paps."

Having made up our mind as to which tree would be the Christmas tree, we decided to extend our walk across the farm and return with the saw. We saw plenty of deer prints in the soft mud and at one place, we saw rabbit fluff where a coyote or owl had taken a rabbit by surprise.

This time of the year I start to look for sheds, dropped deer antlers. But before I could finish my sentence to tell my daughter to keep on the lookout, she finished my thoughts, indicating, "You have told me before, Paps." I know I have, but repetition does not hurt. Walking, we circled the farm. When crossing the new brome, we looked for the tiny sprouts giving us hope for a better stand and a better hay crop this coming year.

Back at the truck we picked up the saw and retraced our steps to our tree. Cutting through the trunk, we carefully laid the tree down, making sure not to break any branches. Then, between the two of us, we carried the tree out, talking as we did. I enjoy walking or working on the farm with offspring. A grandson wants to work with me this summer. That is going to be fun.

At the truck, we slipped the tree in the bed and left the tailgate down so we wouldn't break the top. We called Wheeler, the dog, who jumped in. Driving home, we felt good. We had a good walk and brought a nice tree home. At home, I set the tree in water to allow it to soak up enough moisture to stay soft.

Sunday, I put the tree in the stand and shaped it. It is the shaping which always fools our friends who when they see our tree ask, "Where did you get that tree? It is beautiful."

It is the result of shearing. I do it when the tree is in the stand. I walk slowly around the tree and snip--snip--snip. The result is a beautiful form which then is much easier to decorate. Of course, decorating the tree has stories.

When the children were small, they helped. But I think I might have given too many directions trying to keep the decorations balanced. However, I remember one Christmas where my youngest daughter sat high on my shoulders and leaned far forward to put the top on the tree. I know because there is a photo which shows her serious face concentrating on the task at hand--placing the top!

As I unwrapped the ornaments, many held memories. If nothing else, another year has gone by, a year in which much happened--good and sad. We must hope that the sad is overcome by the real Christmas story. And legends told, both old and new, are part of the cycle when days are short and nights long.

As you decorated your tree, I hope good thoughts were with you. It won't be too long and all the decorations will be taken down once more. Handle them carefully and while the house is all quiet, smile and recall the warmth of the Christmas season.

Merry Christmas to you all!

Life Lessons

The "clanging" we heard.

January 26, 2000

Farmland Gives Way to Urban Sprawl

Flipping pages through a book the other day, I came across an old newspaper clipping written in Dutch from my hometown in the Netherlands. The article was about a farm that had been bought up to facilitate immediate and future town expansion for a new housing development. My father must have sent me the article knowing that it would be of special interest to me. It was because I have very happy memories of being on that farm as a young child. I would play, help with the chores, and spend many vacations there.

I remember following the horse-pulled plow and seeing a nest of young mice being cut up by the plow shears as it pulled through the dark, rich earth. I was too small then to hold the horse-drawn plow, but I learned as they let me steer the horses. I learned to watch horsepower, how to turn the horses pulling the plow around without getting tangled in the traces. I learned when they did get tangled how to untangle them. I learned to trust horses but also that they could kick.

It was on that farm I first saw chemical fertilizer being used. It came in bulk and needed to be mixed which was hard work done with shovels. This was shortly after the war. I remember after a day's work with the horses, the farmer would lift me on the horse, give me the lead rope of the other, and I would take them down the road to the pasture where they would graze during the night. I loved it all--the chores, the seasons, the fragrance of the earth, the hay. I loved the coffee brought to the field--time to take a break. I was only a child but could participate.

I looked for a long time at the photograph of the old farmstead. During the war, I remember a German fighter plane crash landing on a plowed field close to the farm. We were not allowed to get close. The Germans placed guards around it. And I remembered that one day there was a huge pile of hay in the barn. A pile I had not seen there before. When I took a closer look, I saw that there was a car hidden under the hay. I recognized the car as belonging to the local doctor. The Germans wanted the car, just like they wanted and took the good horses, later even the old horses.

The doctor needed to hide his car and he had driven it to this farm. They were willing to help until a permanent hiding place was found. After all, a loose pile of hay was not too good of a place in which to hide a car. It is interesting that a child saw so much, knew so much, and knew not to talk. I was eight, ten, eleven years old; thirteen years old when the war ended.

One day when I walked through the orchard looking for a ripe apple, I noticed the big sow lying in a cool mud puddle against the fence. I don't know what made me do it, but I quietly sneaked up on the resting pig and then quickly straddled it. The sow jumped up and took off. This is just what I wanted, but I had thought the pig would run through the orchard and give me a joy ride. However, the squealing sow had other ideas. While I held on to both ears, the pig ripped along the barbed wire fence. The only problem was, having straddled the pig, my right leg was caught between the pig and the barbed wire fence.

To this day, I still have a wide, white scar on my right knee to remind me of my pig ride. The barbs ripped my knee open and caused a deep cut. In those days, they did not stitch things. I believe all I got was a bandage, iodine, and the necessary shots. All this came back to me when I saw the photo and read the article once more.

The village took the farm. It is all gone, a memory. A new housing development has grown up where I helped with chores, walked, played, and worked as a child. It happens all the time. I was reminded of that fact when I recently came across some statistics of American urban sprawl. The article, "Open-space Development Doubles in U.S. during '90s" by Philip Brasher, states that according to a U.S.A. government report, suburban sprawl has doubled the housing development of farmland, forests, and other open space during the 1990s. Nearly sixteen million acres of land were converted to development between 1992 and 1997.

For some farmers who live near cities, the higher land prices may be a windfall. After a life of working hard, they can retire. However, I wonder how they feel. Around Kansas City, Lawrence, and Topeka I see the old farmhouses, the barns, the fences, no longer needed. The gates have been thrown aside. No livestock to hold any more--roads are put in. Even though an old farmer may be glad that he could sell at a top price, I don't think he wants to talk about it.

And I--I often see in my mind a pig, a big sow lying in the soft, warm mud against a fence, waiting for a young boy.

March 21, 2007

Write Like We Do

Our mailman should be a diplomat. As a matter of fact, he is a diplomat as he did something the other day in such a very sensitive way that I could only laugh and laugh.

We have different people delivering our mail. All are very professional and helpful. At times they have to drive up the long, steep driveway to deliver a package or obtain a signature. When the weather is bad and the driveway is icy, I suggest they drop packages with my neighbor who lives closer to the road or simply shove them into his mailbox as that mailbox is a very large one, built to take packages.

At times, our mailman is a mailwoman with a cheerful personality. However, the day I stood there laughing, the mailman came driving up Marlatt Avenue. I saw him coming as I returned home from a trip to the hardware store. I parked the truck at the bottom of the driveway and crossed the road, making sure no traffic was coming.

It used to be that the mailbox was on my side of the road. But we were told to move it to the other side as the mail route had changed. Over the years, we have moved it a couple of times from south to north, north to south and as I said, presently from south to north again. The only things left on our side are the small tulips that I planted many years ago near the base of the then mailbox. They were the red Apeldoorn tulips, a very strong tulip that keeps coming back even though they are now smaller than they once were. If I gave them some fertilizer, they probably would be large bloomers once more.

Anyway, that morning I walked across the road and stood by the mailbox as the postal truck moved up and slowed down. Having said our "good mornings," the mailman handed me my mail. As he did so, I noticed one letter lying in his lap where he briefly had placed it after pulling it out of the bundle of mail he just handed me.

It seemed he was slightly embarrassed having pulled the one letter out.

I noticed immediately my rather bold, clear handwriting. It was a letter I had written to my son, Guido, and his family in Raleigh, North Carolina. I looked kind of surprised and my mailman could not hide it from me, even if he wanted to. I said, "That is a letter I wrote and mailed the other day. How--how come I got it back?"

The point is, I was not going to be handed it back. My mailman knew, recognized the handwriting as being mine. But I had asked and he answered, "It is your writing. You have to write like we do, more American. The sorting machine cannot read or recognize your writing, especially the numbers."

He said it so gently, not wanting to offend me that I felt warm all over. Here was a friend. I looked at the letter with him and he pointed out some of the differences. For instance, my numbers are written the way I was taught. He pointed out the seven as being different--the downstroke being way too long.

Even though to me, the address was very clear and neatly written, to the sorting machine it was not. It could not sort it. In the past, I had a similar problem with letters addressed to Auburn, Washington state. Then it was the eight which caused a problem. Now I do know that my eights are different. I start them differently and end them differently. I now write my eights very distinctly and it seems to work. Those letters have gone through.

My mailman finished by saying, "I am taking the letter back and will hand sort it." He said, "I know your handwriting. I will take care of it. But it would help if you wrote a little more like us, the American way."

To be told that after 33 years is kind of amazing, humbling. It has intrigued me why we do write certain letters and numbers differently. If you ever wonder, compare the figure eight and you'll see.

I have always admired my great secretaries who managed to read my handwritten notes and letters and type a formal letter or this radio program draft. I must admit when I write fast, my handwriting suffers. But Christy sees the flow of the sentence and she types faster than I can blink my eyes. A few times I have asked her, "What did I write there?" It is true, I have asked.

But the addresses on the envelopes I have always tried--tried hard to be clear, neat, and legible. But it is not enough. I have to learn to write the American way; more like "we do" as my mailman said. A great guy. A great diplomat.

March 11, 2009

Effort!

Manhattan is a university town and several of my friends are colleagues. Some are retired and others are still in the saddle. They are still teaching, doing research, and/or extension. One of them recently forwarded me a report with the comment, "This report may interest you."

It was a report about students' perceptions in regard to being graded and their expectations that effort deserved a good grade, even a B. I read the article a couple of times. Each time, I read that effort--trying hard--deserved a good grade regardless of what was learned. It seemed to say that even though very little was learned or comprehended, if you really tried hard you should get a good grade. What utter, utter nonsense.

I looked at my computer screen and took a deep breath. My friend and colleague who had forwarded that report needed an answer. I knew what he was thinking. He and I would agree. But he was in the saddle and the next morning he would lecture again, standing in front of his students. And he is a good, compassionate man. He would recognize effort, hard work, but what if hard work does not get you anywhere?

I have always been glad that I was not an A student. I have good common sense and determination. Once I set a goal, I go for it. However, if I do not see a clear goal, I don't work that hard. At least I did not in school. I was a dreamer. I mastered courses that interested me. The others I struggled with. Being envious of my smarter buddies who with little effort would make good grades, I had to push--push--study--study--study to comprehend and make a decent grade. Effort was not rewarded, only knowledge, comprehension, and understanding. If effort would have been rewarded, then I should have made all As for those courses I cared about.

To me, effort is like oil in the engine. Without it, most of us get nowhere. The engine freezes up. The danger in rewarding effort is that one can get through school with good passing grades and yet know nothing. Imagine a pilot who passes on effort in his training to learn to fly but does not comprehend or know it. He is given his pilot's license and starts to fly and then crashes. Oh, he put in a great effort. He tried hard--very hard--but did not or could not learn to fly. He was not weeded out.

I am not saying that effort--hard work--should not be recognized. But that recognition should be used to help the student find his or her path where they can master the subject matter to benefit them and their profession. Here, compassion and expertise in guidance are required. Honesty is a requirement.

I will give you an example. Years ago, when I returned to VPI & SU University in Blacksburg, Virginia, I was an older student. I had three children, later four. I held an Australian degree H.D.A. but wanted to pursue a B.S., M.S., and later decided on a Ph.D. Landscape and Environmental Horticulture was my choice. I took tough graduate courses and worked hard. It was serious business and my wife encouraged me and the kids loved me. So did my wife, by the way. We ate a lot of peanut butter and jelly sandwiches.

Anyway, earlier I had taken a course in geology which I liked and saw as a foundation to the field of landscape architecture. I was studying the skin of the earth--the land formation and landscape. It made sense. When I saw an advanced graduate course on Appalachian geology, I knew that was a course to take. However, I was not a geology student. But never fear. I walked to the professor's office, knocked on the door, and explained my case and interest. He looked me over, apparently liked what he saw, and let me into that class. I was happy as a lark.

The course was a graduate class that had no tests, only one final exam. The practical side of the course took us all over Appalachia. I enjoyed it. I learned. I saw connections. The landscape of the old worn mountains started to make sense to me and I understood where I was.

Then the day of the final exam came. Several hours of quiet while we students concentrated on writing, calculating, and sharing what we had learned on the paper in front of us. I was stumped. I could not answer to the depth which was required. I thought to myself, "You fool, this is the moment of reality." Mind you, I had studied. But remember, I was not a graduate student in geology. I had no foundation beyond enthusiasm, hard work, class participation--yes; knowledge--no.

A few days later I received my exam back. I got a B. I went to the professor and said, "I do not deserve a B. No way."

He looked at me and said, "You earned that B." Adding, "As far as learning, you probably learned more than any of the other students in the class. As a line on a graph, your learning went straight up. You added to the class by asking questions."

"However," he added, "if they ask me for a petrologist--a geologist--I will not recommend you!"

June 7, 1995

Farm Stress

They have seen it all come and go and come again.

As I stepped out of my truck, it started to drizzle a slow rain--just enough to keep everything from drying out. And drying out the soil is what was really needed. The gravel driveway was still muddy from a shower that had dumped more than an inch earlier that morning. More of what farmers don't need right now.

I walked over to the men working on the swather. They had parked the machine in front of the shed under a low tree with spreading branches. Only half of the swather was under the umbrella of branches. The other half stood in the open and standing there I got wet. A slow soak I call "dry rain." They were fixing a bearing that had given them trouble. With some disgust, my farmer friend held up the paper bag and said, "Just in the bottom of this bag is $75.00 worth of new parts and look how little it is. Yet it costs $75.00 and that is a small repair."

Some of the new parts looked shiny red as compared to the faded red on the much used and dented swather. But I know that when the swather is pulled into the field it will do the job of mowing and rolling the fresh-cut hay. Because even though the machine is not new, they keep it in good running condition, greased and oiled. Keeping the farm machines up and making sure they work properly is the only way for most to stay in business.

There under the tree, while working on the swather, we talked farming and profit. The alfalfa needed to be cut but it was too wet, and more rain was predicted. Soon the brome would need to be cut, but the soil better dry out. In a few more weeks, the wheat harvest would start but we better have some hot, drying weather to make the wheat ripen. Beef cattle prices were way down. It did not look good. With all the rain there was enough grass and the cattle looked good, but some warm sunshine might even help there.

Some extended sunshine would surely help dry out the unplanted fields so that even though too late for corn, milo or beans might still be planted. The ground has to dry out before the equipment can get into the fields. Working the soil when wet will do lots of damage to the soil structure which might take years to fix.

As the slow drizzle became a steady rain, I moved under the umbrella of branches where it was dry while my friend kept working on the swather, ignoring the big raindrops. It is that stoicism in our farmers which ultimately, I think, makes them successful, or if not successful accept their chosen profession with all its risks and hardships. Farming is a way of life and once chosen, it is very hard to leave behind.

Yes, some have been forced off the land, either from bad luck or bad management.

It is the bad luck that frustrates me, and I admire those farmers whom I have met who keep their dignity under all the bad luck which natures hands them. Bad luck as simple as more rain, more rain, more rain one year and extended drought other years.

Many years ago, I tried to get a tractor started which I desperately needed to get a farm job done. The tractor would not start. Seldom in my life have I been so mad at just a thing--a large metal thing on rubber tires. I tried everything--it never started. My wife, who heard me rant and rave, walked away slowly, unable to help me but also unable to watch my anger at a machine that refused to start.

Over many years, I have learned to control my temper. Admittedly, it took a long time. But I still feel the agony, the hopelessness, when luck seems to be against you and it comes pressing on you with all the consequences.

It is the ability to handle the pressure which I admire in the farmers, anybody who grows crops for a profit and is dependent on nature. The older farmers seem to be the mellowest. They have learned. They have seen it all come and go and come again.

As you drive the countryside this summer on vacation or just for a weekend outing to enjoy the scenery--the rural scenery--remember, when you grab your camera and step outside the car to take that colorful photo of an alfalfa field in bloom, it should already have been cut. For the farmer's sake, when you see a wheat field coloring, hope for some dry, hot weather. When you see a sunflower field, know he got it sown after all. Later in the season when you see soybeans ripening, hope that an early frost does not get it.

I sometimes think everyone should grow something, anything, which could be lost in an instant. It can make you rather humble with a different outlook on life.

August 15, 2012

The Marble

The old jacket hung between my clean shirts and pants. It was often pushed aside as I selected a clean shirt to wear. I knew it was there and when I say old, I mean old! However, as jean jackets go, old may be the new "in" with its holes and tattered threads--some may say that is fashionable.

When I looked at the fabric, I was amazed at how strong denim material is. Threads still hang together even though you can poke your finger through a hole. All of the buttons are still on the jacket, and they close the jacket over the buttonholes. Although there are a couple that constantly need re-buttoning as they have a tendency to slip out.

But I wasn't wearing that jacket anymore, it was retired, faded, having been well worn. Where many old worn and torn jean jackets are ultimately tossed, this one was saved and even though I did not look at it often, I knew it was there and it was treasured. Because the jacket had a story to tell, an old story, the story of a 26-year-old. But the jacket is even older than the story--much older. The jacket was well worn already then, my favorite. I loved it.

One day long ago, I had my jean jacket on when my youngest son, Martien, and I walked the path on the hill behind the house. Martien was sad. All his siblings had left the house and his oldest sister was moving to Ohio with her young family. As we walked the hill, Martien, nine years old then, looked up at me and expressed his sadness with all the changes around him. It might have been even more sad because young as he was, he was an uncle to Joseph, my daughter's oldest son who would be dropped off by one of his parents early in the morning. All those close connections would come to an end.

I understood Martien's sadness, but I also knew such was life. At least the siblings did not move overseas. I explained this to Martien and I pointed out the four directions: north--south--east--and west. I explained where Ohio would be, roughly which direction. Then I picked up a small stone and told him, "Now look when I toss this stone in the direction of Ohio. It will roll a little and that is how it should be."

He watched me toss the stone toward the direction of Ohio and the stone rolled down the hill. He stood there. I could see him thinking. Then he put his hand in his pocket and pulled out a marble. He looked seriously at me and asked me to stoop down and he slipped the marble in the left breast pocket of my jean jacket and closed the flap. Then he said, "I don't want to be tossed yet. I still want to be close to you. Keep me there in your pocket."

For years the marble stayed in my pocket. One day I asked my wife to sew the pocket shut so I could never lose the marble while wearing the old favorite jacket. And then years later, after another washing, the faded jacket which had aired on the line, bleached by the sun and wind as only jean jackets do, was taken off the line at the end of the day and hung on a hanger next to my shirts. It hung there for years like an old, loved horse put out to pasture. Once in a while, I touched it as I moved shirts to check the pocket. The marble was still there, safely sewn in, holding memories of Martien, now 35 years old.

Over the years as I pushed the old jacket aside, I came up with an idea to celebrate this jacket. I would never wear it again. My wife would not let me. And honestly, I had to agree. Although taking the jacket off the hanger, I slipped it on and it still fit and felt good. It fit like a glove, feeling like only an old jean jacket can feel--very, very trusted. Standing in the sun, a photo was taken. That was for posterity!

Then we took the jacket and folded it in a way that the pocket with the marble became the centerpiece. We laid it out carefully, draped on a background of weather-beaten, green-gray boards. Three boards, 50 inches long, screwed on to cross boards making it 24 inches wide. I left spacing between the boards. The top half I framed with an old window salvaged from the old, fallen house on the farm. It was all left rough and rustic. The lower sash of the window was gone.

I carefully stapled the jacket down, making sure it was draped with the pocket on top. One sleeve holds a braided rein which Martien had for his pony Poptart. I looped it around and behind the sleeve, nailing it in place. To the right, above the jacket, I attached two horseshoes with special nails. Those shoes were worn by Lips when he and Martien won a Grand Prix.

Below the jacket, I attached with clip unlatched, the tiniest halter worn by Thunder who was one of Poptart's foals. Thunder grew into a great pony who loved to jump. Martien was the very first one who eased himself onto his back while he was standing freely next to the gate. "He did it!" I found a photo button of Thunder as a young foal and attached that.

To complete the picture, my daughter and I drove out to the farm one morning and found the horses, Poptart and her grown daughter, Misty. While Talina fed them a handful of grain, I cut the ends of their tails. One bay and one gray. Back home I bundled each and using staples, attached them below the sleeve. That is enough memories. Oh yes, this story and the photo of me in my old jean jacket for the last time will go inside the jacket.

Now it's ready to go to Martien. It is part of his life--my memories.

August 27, 1997

The Model Tractor

It has the red color and yellow letters. The Massey-Harris model tractor stands above my computer. It just fits there, front wheels turned as if it is turning to make another round before calling it quits--or before the job is done. There is a difference. Sometimes the job is so big you just cannot finish it in a day and because the weather is stable, you decide to call it a day and finish tomorrow. It all depends on the job. Some fieldwork you have to quit because conditions are not right, the field is too wet, or the last corner which lays a little bit lower has not dried out yet.

It is like Bud asked me last Saturday, "You think we can work the fields, or are they still too wet?"

We had plowed the stubble under and whatever rain we had soaked in nicely. Yes, some parts of the fields could be disked now, but the low parts would still be somewhat wet, and we try not to compact the soil or work the land when wet. We try not to--but sometimes you have no choice.

The model tractor, a red Massey-Harris, is a 55 model built on a one-sixteenth scale. It is exactly as I remember--even the yellow letters. I expect my youngest grandson would love to play with it. However, I did not buy it for him or anyone else for that matter. I bought it solely for sentimental reasons.

When I saw the little tractor, I recognized the model instantly as a tractor I had worked with when living in Australia. Just looking at that model, it all came back to me--a young guy on a huge farm close to West Wyalong, a town in inland New South Wales, Australia. I was plowing wheat stubble and it was so hot--bloody hot--that when I plowed into the wind the long straight furrows, I had to stop halfway in the field to let the engine cool down. In those days, we didn't have air-conditioned cabs. It was just the wind that kept you cool, except this was hot--like a desert wind.

It all came back to me just seeing this model tractor and suddenly I had the urge to own it, so I could look at it and remember 42 years ago. The model tractor was to be auctioned off for a good cause.

My secretary, Christy, saw the smile in my eyes and after I told her, she asked, "Do you want me to bid on it for you?"

I said, "Yes," since I could not be there.

To my happy surprise, she came back after the auction and said, "You bought yourself a tractor!"

It is now standing above my computer and it is rather interesting how farming practices have changed over 45 plus years. I just read on my computer how today it is possible for farmers to farm their fields and collect instant data from satellites. That data will relay soil conduction, crop conditions, etc. which then can help farmers make the kind of decisions needed for micro-management of field and cropland to optimize income.

Of course, a good farmer knows his land, and the old saying "The best fertilizer on a farm is the farmer's footsteps" still holds true. But satellite imaging has made it now possible to do in the office what footsteps used to do in the field.

Walking over my cultivated land I know where the soil is shallow and where it is deep. I know where the rocky, dry spots are and where the wet spots are. At times, I may direct more fertilizer to be put on the upland to help stimulate crop growth. I still work with common sense using footsteps.

Of course, my acreage is not so big, only 435 acres. And don't get me wrong, I am duly impressed with all the technical advances which can be bought and supposedly improve farming. However, I am glad when I watch Bud scoop up the soil and see him let it slip through his fingers and hear him say, "The soil is getting better, we are doing the right thing."

We are doing the right thing with the management decisions we make, the equipment we have, and do not forget, the opportunity which I sometimes call "the cards" Mother Nature deals us.

I never thought I would collect tractor models and really I'm not, but this 55 Massey-Harris I just had to have. It was part of my life when I learned what it was like to be lonely and the best antidote was work, work, and work!

September 23, 2009

The Half Full Cup

Last week I had a day where I did not know if my cup was half full or half empty. Of course, we all know that it makes a difference in how you look at things.

The weeks before, my wife had gone through two operations which had caused me to ask the good nurses, "Can we request our own room?" The fact that all went well was surely reason to call the cup half full. But that was earlier.

To reflect on the bad day last week, it started when I drove to the farm to check on the horses. I had not been there for a while, having been--or tried to be--a good nurse. That job was made much easier with the care given by my two daughters. I said to them, "I can't wait until I get sick someday and you two take good care of me the way you take care of your mom."

In unison, they said, "Oh, no. When you get ill, we'll call the boys." Thank you!

Anyway, on the farm I found the horses in good shape. There are only two now. So they get more affection. I looked them over, wiped some sitting flies off their backs, and checked their feet.

Seed heads were waving in the breeze around them. There should be a very good burn next spring. But first I must walk the hills with a chainsaw cutting the bigger cedars down. I don't understand how they survived earlier fires. They are a weed and must go. It is work.

When I crossed the brome pastures, I noticed fresh cow piles and cow tracks across the hay fields. Looking further, I saw the cows--red Aberdeen Angus lying contently in the shade. I knew whose cows they were and seeing the trails, cow piles, and grazing patterns, I knew that they had been on the land—trespassing--for quite some time. They looked fat and content with their shiny coats and very sassy. I saw no need to disturb them, so I let them lay in the shade.

Now my cup was less than half full. I continued walking to the shed and tried starting the tractor. As before, it cranked over but would not start. I drained fuel from the carburetor and crossed my fingers, toes, arms, and anything else I had two of. The tractor cranked but did not start. No--ppptt--prrr--pttt followed by the joyful, easy rumble only a John Deere can make.

I turned around disgusted and noticed where rubbing cows had knocked a pole down. True, when it itches, you have to scratch. But I needed a running tractor. I needed to mow brush and weeds and clean trails. As I looked around, I really noticed how bad I needed a running tractor. Surely it can't be a major starting problem as I had recently moved it from the open field. I understand why farmers keep several tractors.

My cup was now way below half full. Driving home I was disgusted. Except for the horses, I had not accomplished anything and realized that between doing things at home, the yard, etc., I was getting behind. Besides, I knew the tractor would start if only it was used more regularly. I was just about reaching the bottom of my cup. It was getting bitter. I saw some grinds.

When my grandson drove up with his big 4x4 Ford truck, I decided to ask if he could help me move some big poles on the farm. A job I had wanted to do with the tractor. He immediately said, "I'll be glad to help, Baba."

When asked about when, he said, "Why not now--this evening?" I jumped on it and after throwing a chain in the back of his truck, we drove off. His dog, a beautiful German shepherd, was sitting in the back watching the world go by.

The trip to the farm was a pleasure. The big diesel engine hummed along. The evening sky was peaceful. At the farm, we had a beautiful sight. Right on the front pasture was a postcard scene of 26 cows and calves and a very happy bull. They were content, happily grazing the very lush bromegrass. The eighteen-acre meadow and the other 70 plus acres of good green brome had put meat on them over the last weeks. They looked good. Except they were not my cows! Nevertheless, I did enjoy the postcard scene very much.

When we hauled the poles in, the cows came up to investigate and just stood there in the way. Indeed, a very quiet and contented herd--very well managed. But as I said, they are not my cows. My grandson suggested we move them back to where they belonged. I said, "No, let them be. They'll soon lie down after another full day of excellent grazing."

They may even drift home and pretend they never left the home pasture, making the owner believe everything is as it should be. I looked over the small herd of contented beasts. It was a beautiful sight. My cup was starting to fill again.

When we drove out, I closed the gate. The cows would not go anywhere but perhaps home to come back the next day. We drove by my neighbor's house and I told him where the cows were so he would not have to search for them. As a matter of fact, I had smelled them in the morning when they had been lying down. And I really smelled them when the chain we used to pull the poles went through a soft, green, well-spread cow pile. Just as you want it, indicating good nutrition--good pasture--good digestion. Yes, I cleaned the chain. I pulled it through some clean grass.

When I got home, my cup was more than half full again. Lucky me.

October 14, 2009

The Rusted Nail

A small, rusted nail hangs from a thin fishing line on the mirror of my new pickup truck. The truck is a result of the Cash for Clunkers program. It is quite a luxury to drive a new truck, expecting and knowing that everything is new, not worn, and that everything works.

It is no reflection on Old Red. No way. What is true is that the cab of Old Red, the trade-in, was kind of dusty and filled with stuff. Stuff any farm truck needs to have because you never know when you might need it. I am slowly adding that kind of stuff to the new pickup. Slowly. For now, the cab is still clean and organized. All the lights work. The only thing you may call old in the new cab is the tiny, rusted nail with a square head hanging on that thin thread under the rearview mirror.

People who have traveled with me have looked at it with a vague expression like, "What the heck is that?" It is only a two-inch nail and very rusted. It does have a sharp point. It is a nasty looking nail.

When they ask, I reply, "Do you know the story of the sword of Damocles?"

So far, no one has said, "Yes, I know that story."

So, I tell them the story or the legend. It is an old story and today, sometimes a situation is told with the use of the expression "It is a sword of Damocles." Meaning, more or less, "it may backfire" or "things are not what they look like" or "be aware, you do not know what the future holds--don't ask for too much."

Here is the story of the Sword of Damocles. The story is an old Greek legend, well before Christ, lost in the history of Sicily. Later, a Roman historian heard or read about it and it was passed on into European culture. It is a moral story and no question, that is why the old nail is dangling in my truck cab.

Damocles was a courtier, an attendant in the court of King Dionysius II of Syracuse, Sicily. Damocles was always flattering the king. The king got sick of it and invited Damocles to a banquet put on especially for him. The king threw his gown over Damocles' shoulders and invited him to sit on the throne and enjoy the banquet. All the attention was now paid to him.

As Damocles ate, drank, and laughed, he looked around and thought that being king was *the* thing. Then, his eyes suddenly caught sight of a sharp, shiny sword hanging on a single horsehair right above the throne--right above his head. In that moment when he noticed the sharp, pointed sword dangling above his head, Damocles no longer wanted to be king. He lost all taste for the fine foods and wine. He asked to be allowed to leave. He wanted to get away from underneath that sword. He no longer thought it would be fortunate to be the king. King Dionysius had successfully conveyed the sense of constant fear great men live under.

Today, the story of the Sword of Damocles is often used in stories, films, television series, video games, and even music. It denotes the sense of foreboding often sensed in a precarious situation, especially one in which the onset of tragedy is restrained only by a delicate trigger of chance.

And it is the "delicate trigger of chance" which made me save the rusty nail after I discovered it. The truck's informative dashboard already had told me for a few days, "Your left front tire is low on air and needs attention."

Twice I had put air in the tire when the tire pressure was indeed low. But I could not believe or accept I could have a flat. After all, I had a new truck with four new tires. The odometer showed only 2,300 miles. The rubber had only just rolled the pavement and yes, the gravel roads.

Yes, I had driven up to my neighbor's house on his farm. It is a big house which had been moved to this new location several years ago by expert whole house movers. I have always admired that precarious job. It was done so well that it seems the big house was built there--from the bottom up. But could it be that in the move, the old house had lost a nail? A long-lost nail no longer needed and now lying in the driveway, rusting further, and waiting to be picked up?

My tire picked it up. It punctured the wall. A new tire punctured, and they could not fix it because the nail had penetrated the wall. I scratched my head. What now? The solution I saw was to get better tires under the truck. I opted for two ten-plies on the back to help me carry a load. That gave me a spare and a new tire, the old spare to replace the worn front tires whenever that day comes. Hopefully, it will be several years from now. Those tires are only 36,000-mile tires, being typically cheaper tires under a new truck, they may not last that long anyway. A rock in the creek may puncture them while crossing. Anything is possible. There always is that chance. A new truck can have a flat only one mile away from the dealership. It is that chance that made me pull the nail out of the tire, the new tire, no longer good to use.

I put the nail in my pocket. When I got home, I looked for a piece of fishing line. I did not use a single horsehair. I want to keep and hold on to the nail as a reminder. Yes, you have a new truck. Enjoy it but be aware. Anything can happen. Such is life!

November 5, 2003

"If You Wait Long Enough--It'll all Come Back"

Farm buildings are like tools. They have a function and a history, a cultural history like the agricultural landscape. However, it can change.

Sometimes thoughts and ideas come too quickly and become overwhelming. Last week, my wife returned from another visit to the Netherlands and as nearly always, brought back a few books to add to my library. Two neat books came out of the suitcases. Both are Dutch. However, what makes them interesting is that their subject matter is the same as an English book I recently read. All three books treat similar concerns and look for similar solutions to the rapid changes in agriculture.

Recently, I came across once more the idea of the "Buffalo Commons" brought forward by Frank Popper and his wife, Deborah, who suggested in the late 1980s that most of Kansas and other Plains states ought to be turned back into a buffalo range. Former Kansas Governor Hayden reflected on the issue during a talk in 2002 given in Manhattan. Again, I saw the idea while reading *Home on the Range: A Century on the High Plains* by James Dickerson. And even more recently, it keeps popping up on the web and is forwarded to me by well-intended friends.

There is an upcoming talk by Wes Jackson, President of The Land Institute, titled "The Necessary End of Agriculture as We Know It." I made a note of it on my calendar and plan to go and listen. I know that my small acreage has no major impact on the agricultural economy and was reminded by a soil conservationist, "Gus, your farm will go the way of the buffalo." It is only a reminder of being prepared for what will come, the rapid change of agriculture as we know it--actually, a continuous change.

The Dutch books fascinated me as both books show and interpret the tremendous and rapid change in Dutch agriculture, the change in the culture of agriculture.

For a starter, remember that the Netherlands is only one-sixth the size of Kansas. So visualize how large that would be--not very large at all. It would be like cutting a loaf of bread in six equal parts and only one part, one-sixth, is the Netherlands--with a population of nearly seventeen million people.

At one time, the Netherlands had very strong agriculture. A culture of agriculture expressed in the beautiful farmsteads with strong regional architecture reflecting the kind of farming practices--from dairy with large stalls to grain production with large sheds. In 1950, there were still more than 400,000 production farms. By the year 2000, there were around 100,000 and as time goes by, fewer and fewer farms remain.

The land is sold, joining and enlarging other farms. There is road construction, urban expansions, all of these familiar reasons. However, the historic farm buildings left standing are of concern. And the book I was given digs into the issue of how to save these farm structures and homes.

Some very, very interesting solutions have been found and it reminded me of the program we have called "Barn Again!" as a play on "born again." The historic and cultural value of the buildings and the core of the farmstead were recognized and an effort was made to save the structures and that part of Dutch culture. There are structures built in the 1600-1700-1800s with mighty oak beams and post structures all held together with pegs. The barns are no longer functional. However, they are beautiful when restored to their old glory and given a new purpose. Their function may be as diverse as a restaurant and conference center. Remember, the Netherlands is much smaller with a denser population.

The detailed drawings fascinated me, being very familiar with the type of buildings both here and there.

This past weekend, a friend of mine showed me a Dutch map of a region we are both familiar with, having bought the map on E-Bay. It is an antique map from 1600 showing the landscape--forests, swamps, old roads, old towns, and old communities in the Netherlands. If you lay that map over recent maps, you will see how land use--agriculture--has changed the landscape. So-called wasteland was brought into production, forest land was cleared, and swamps were drained. Towns and cities expanded, roads and more roads were built as the population grew. However, as agriculture has become less important, nature in the densely populated country is needed more and more.

There are big plans to take some of the reclaimed terrain, now agricultural land, and revert it back to its natural state of swamp, forest, and even give some back to the sea. It intrigues me.

The first time I heard about the "Buffalo Commons" I could not believe it. However, it keeps popping up as an idea. People keep talking about it. The Dutch, as well as other Western countries, have similar thoughts: return land to what it once was.

When you read *Home on the Range: A century on the High Plains*, and you realize how stubborn and resilient the people who farmed the land are or were, you know they are special people.

But I will never forget what the old retired and bent-over farmer said when my mother started to plant trees around the old Dutch farmstead where he grew up as a boy. He said, "Then to think that my father and I grubbed and grubbed trees and drained this wetland so we could farm, eke out a living."

I wish I could see what our landscape will look like 50-70-100 years from now. Will there be buffalo roaming once more?

December 2, 2009

Raking Leaves

It was a great day with a cool wind blowing from the north. The day before, a friend drove through a blizzard on her way back to Denver but here in Manhattan, we enjoyed one more unseasonably warm day. By evening it started to change and a thunderstorm rolled through, hence the cooler temperatures the next morning.

I reasoned that with the rain that fell during the evening and early night, raking leaves might be a pleasure. The next day, the sun was out which did indeed make working a pleasure in the cool breeze. Also, with the wind coming from the north, it helped my raking pattern as leaves would blow with the raking.

Of all my garden tools, there is only one tool I associate with my father and that is the leaf rake. It is a good one. Actually, I have two, but one is much better than the other and that is the one I associate with my dad. It was when we still lived in Blacksburg, Virginia, and my parents visited us. It was fall and the maple leaves in front of the old farmhouse had dropped and were lying thick on the little grass there was under the big maples.

I do not remember why he started raking. No one asked him to and I am sure that until he wanted to rake leaves I did not have a leaf rake. I must have bought two so that I could help him and we could work together. I know he enjoyed the job. It was not a mindless job to him because his mind worked all the time. I am sure it was relaxing and he could, as he might say, organize his thoughts.

As piles grew, the kids, his grandkids, tumbled in and over the piles of leaves, scattering them. I don't know if he minded or simply watched them play as he steadily worked and raked. He did not work himself into a sweat. He was enjoying himself.

My parents visited us a couple of times when we lived out east. I regret that they did not come to Kansas when we moved here. By then, they had aged and the long trip had become too much.

But in my mind, I still see him rake--rake--rake the colorful leaves, enjoying being outside with the children and his own thoughts.

I treasure that old, but still very good, leaf rake. It was a good 35 years ago, probably 38, that he used it. But that is the only garden tool he was comfortable with. As we grew up, the garden tools were my mother's. Someone once broke a tooth off her special garden fork. She had it carefully repaired and it was like new. But she always remembered someone had been rough with it, and we kids knew not to use the garden fork. In her garden, my mother did wonders, creating and growing plants--a colorful garden.

As I raked my fallen leaves, the wind picked them up and played with them. However, I kept them together, piling them and hauling them off to the compost pile. To haul them, I used an old bedsheet which I spread out in front of the pile, flipping the leaves onto it. When full, I grabbed the four corners, held them firmly together, and carried the bundle over my shoulders. When I saw my shadow moving across the lawn, it looked like Santa Claus going on his round on Christmas Eve, a bundle of wrapped up leaves--the bag with presents.

I kept my eyes on the ground, pulling the leaves together. By looking at the leaves, I knew exactly under which tree I was working. I remembered the history of that tree, who gave it to me, or when and where I had bought it.

There were the large leaves of the London plane tree. That tree was a small tree when we moved there. It stood very lonely beside the garage. Now it is surrounded by other trees. Every now and then, I picked up the smaller--but still red--leaves of the ornamental pear. The White oak I bought and is growing well. The Bur oaks are really thriving. The Kentucky coffeetree was a gift from Professor Ray Keen who always had a small nursery growing things. The leaves of the Golden rain tree, the apple trees, the locusts, and the true Chinese elm are all small and tend to slip through the leaf rake and stay behind. However, I wanted them all and I got them, working the rake vigorously.

As I pulled the leaves away, the green lawn came back. And that is the only reason I raked the leaves. I am trying to grow a better lawn and even though I believe in mulching, burying a lawn under thick leaves will smother it. I finished the job. Now all I need to do is wait for a day with no wind and go over it one more time, raking the remaining leaves together and add them to the compost pile.

Which reminds me of when my brother and I were little, we had to rake leaves in our garden at home and put them on a compost pile. But then when spring came, we had to take the composted leaves and put them back in the garden. My brother and I always argued with our mother that we should leave them where they fell. It would save us much hard labor. But oh, no! We had to do the job and do it well. Amazing what you learn from your parents.

I cleaned the rake and put it back where it belongs. It is a good rake!

December 29, 1993

Happy New Year

I changed my plans when my son said, "Dad, it is going to snow tomorrow."

This past Monday, my youngest son looked worried and somewhat hesitant when he came back from town. "They are talking about a chance for snow tomorrow--a 40 percent chance," he said. "Do you think you could take me and the horses to Kansas City today?"

Way in the back of my mind I had considered taking them on Monday instead of as planned on Tuesday. The reason being exactly what he said, a chance for snow and slick roads--road conditions I do not like when pulling a horse trailer. However, earlier I had made the decision to go on Tuesday and work together on Monday to get jobs done where two men make the difference.

So, I did not immediately jump up when he asked if I could take him today, Monday. But I know what it meant to him to get safely and timely to Kansas City. So I said, "Okay, let's get ready and go."

By the time we loaded the horses, it was early afternoon and bitterly cold. I was glad I had a closed trailer; even so, the thoroughbred was blanketed. But the pony with its thick coat was comfortable as warm toast.

As we drove out, I reminded my son that this was the day he was going to help me and I said, "It had better snow tomorrow, otherwise I will feel 'had'."

He chuckled and said, "It is too cold to work outside anyway."

I partly agreed but reminded him that the jobs needed to get done and I expected the weather would now settle for winter. The warm shirt sleeve weather from Sunday I did not expect to see again until next spring. Of course, you never know!

Anyway, the decision was made and we were on our way with both horses settling in for the ride. We beat the Kansas City rush hour traffic and once at the stable, unloaded the horses, stabled, and cared for them.

Looking at the sky, I thought I noticed some lifting of the heavy, gray clouds. I saw less chance of snow. I delivered another punch line to my son saying, "If there is no snow, you owe me two days of hard work."

He gave me a big hug and added, "Thanks for taking me, Dad."

My wife and I drove back home in the early evening. I expected to see the first snow flurries. We noticed that the temperature had risen several degrees. By the time we got home and parked the truck, I noticed a watery moon overhead, peeking through the clouds. It looked like it was clearing, and I knew then that the weatherman had favored my son.

When I woke up early Tuesday morning, the first thing I looked for was if it had snowed--no, it had not! I just grinned. I knew my son would be thinking the same thing and he too would laugh. I also knew he would be busy riding horses the following days, young horses that need training. He loves the work.

But I had lost my help and as the day went by, the weather cleared--there was even some blue sky. From our hill, I could hear the airplanes land and take off, taking many Wildcat fans to Tucson, Arizona, for the Copper Bowl. I thought for a moment, if the weather had been nasty, these football fans might not have taken off.

As the day passed, the temperature seemed to climb. Working outside was comfortable. With daylight fading, the sun set with a brilliant orange light peeking from underneath the evening clouds while in the east, a huge, full yellow moon rose from the hills. In the east, the sky was clear and cold with this gorgeous moon slowly rising. The last light of the day turned the grass on the hills a soft orange as the sun set further and the last rays faded.

For a while, I stood at the top of our hill and watched the moon. Below me, the lights of Manhattan twinkled with remaining Christmas lights, reminding me of the holiday season. In a few more days, 1993 will be history and we will make plans for 1994. As I watched the moon, I realized there was a lesson to be learned from this expected snow day that forced me to change my plans.

I better not make too many rigid plans for 1994. Instead, take it as it comes and enjoy each day and above all, look for the moon! I know it is easier said than done, but I am going to try--Happy New Year to you!

Home

Barn with a straight roof.

March 14, 2001

Digging Up Your Lawn by Moonlight

They say every cloud has a silver lining, but sometimes you have to look for it. This past Friday our son, Ian, visited us with the two grandsons while his wife visited her mother. Ian drove up on Thursday evening and the kids quickly settled in--running outside, inside, outside, and visiting their aunt and cousins who live in Manhattan. It was to be a fun-filled weekend, and it was.

Friday morning, Ian and I cleaned a path for the trencher to lay pipes that will soon connect our house into the new manhole, which is part of the new sewer line construction, the result of being annexed.

Ian, the engineer, and I, the landscape specialist, came to a consensus about where to put the line through the trees. Trees we planted years ago to give us wind protection and to provide habitat for all kinds of wildlife. Keeping in mind that gravity will do the job, our basement should never flood again being that the house is on a hill and we used the principle my father-in-law taught which is: the shortest distance between two points is a straight line. So gravity and a straight line should solve any future flooding problem. Of course, we were not hooked up yet. All we did was prepare the site, haul off the branches, and make a huge brush pile which I plan to burn next snowfall. It'll snow some more, don't worry.

After a morning of hard work and a job well done, Ian and I decided to check out the farm in the afternoon and give the dog a run. It was about five o'clock when we came home. It had been a beautiful afternoon. Things were greening up and if you looked down, brome was sprouting and ponds were running. All in all, a great day!

Stepping into the house, everything was very quiet. The laughter and the chit chat of the kids were gone. No voice that asked, "You guys ready for something to drink?" No, there was no one home.

There was a note on the table which I expected to read, "We'll be back soon." No, the note said, "Sorry, but 45 minutes after you guys left, I noticed water in the basement. I moved furniture then went to Talina's and took supper with me." Everybody was supposed to eat at our house.

I said, "Wow," then something else, and went downstairs. Ian followed closely, neither of us liking what we saw. I expressed my thoughts to Ian and said, "Could this problem not have waited a week to happen?" By that time, I would have been hooked up to the city wastewater system. Oh no, it just had to happen now. It could not wait--oh, shucks!

Talk about a silver lining, luckily it was clean water. I started vacuuming but that did not solve the problem. After looking, we found it. Luckily it was clean water, remember the silver lining, but the carpets were soaked. I knew that the septic tank was full and needed pumping, but during the inspection I had been advised to wait until the actual hookup because it would have to be pumped then. And being a Dutchman, I hate to throw money away.

Having surveyed the problem in the house, we realized that the solution lay outside. We grabbed spades and dug out the lid of the septic tank. Yes, the tank was full--full--full. At one time, I knew where the distribution box was located, but I could not find my diagrams. By now, it was getting close to eight o'clock and even if I did not feel like eating, I knew we better clean up and go eat--which we did. My daughter quickly filled two plates and everybody was kind of gloomy. I know I was very upset that this had happened, but no one could help it.

Later, my wife said, "Why are you mad at me?" and I said, "Honey, I am not mad at you--I am just mad." I chuckled when much later my son asked, "How come men can't just be mad?" He apparently has run into that same problem that wives take things personally when you are mad about something.

It was dark when we got home and I was ready to call it a day. The children stayed with their aunt and Ian left to visit. My wife went to bed while I once more looked at the project at hand: solving the water problem in the basement. Standing on the front lawn, I realized how light it was with a full moon. So, I got ahold of the shovel and once more started digging. I looked for the distribution box.

Ian came back at twelve o'clock and I watched him switch off the lights in the house and turn in for a good night's sleep. He did not know I was outside. I kept digging. It was now after midnight and I thought, "I am glad we live alone on this hill or my neighbors would think I was crazy."

The highlight of the evening was when my wife called out, "What are you doing?" Apparently, my answer was not very nice, but here I was digging up my lawn in the moonlight to solve our water problem! In a little while, Ian came out of the house and said, "I thought you were in bed."

I said, "No, I am digging. I need to solve this problem."

He put on his boots and started digging with me. We talked the problem over and decided to look for the connecting pipe and burst it to let the liquid flow out of the full septic tank. We found the glazed red clay pipe, dug a pit around it and with the crowbar, one--two--three, cracked it. We could see the flow in the moonlight and watched the level go down in the tank. We unearthed the second lid so we had a clearer picture. Now we had tried everything. It was two-thirty, time to go to bed.

So, what is the silver lining? It could have been worse, bad as it was, especially the timing. One more week and it all could have gone downhill through a new pipe. It could have been snowing. The ground could have been frozen solid. I could have been all by myself. Ian and I working together could laugh. At one time, geese flew over. Coyotes were yipping, no--laughing at us. When we took a break from digging, Ian said, "You know, Dad, I feel like a grave robber working by moonlight."

Perhaps the biggest silver lining was standing there with your son, creating another memory!

May 17, 2000

You Can Come Home

There is a saying, "You can't go home," but I have found that it is not true. You can go home. However, you cannot stay!

I had not been back to the Netherlands for eight years. My wife goes regularly, her mother is still alive and I want them to enjoy frequent visits. It was hard enough when she left home as an eighteen-year-old to follow me to Australia. That was in 1955. Over the years, we have gone back many times with the children while they were young. Later, some of the children continued to go on their own. However, as years have gone by, the older generation has died off--few are left. It is now the younger generation, the cousins, who are establishing contacts.

But I had not been back for eight years. To be honest, I was somewhat afraid of the emotional drag it might give me. My parents died years ago. It was the "old country" that made me go and the wish to attend the 55-year memorial and celebration of the liberation of the end of World War II. In brief, I did decide to go, partly for business with some time added on to show my daughter, her husband, and their two sons--eighteen and sixteen years old--the places where Oma and Baba grew up and lived before they left the Netherlands. Going first to Australia, then in 1959, to the U.S.A., and finally for the last 26 years, to Kansas.

But I was afraid that going back might cause me pain and renew the old question, "Why did we go--why did we leave?"

It was fun to show the kids the places where we grew up. For my wife, the city of Utrecht is home, with its cathedral and large Dom Tower. They climbed all of the 465 steps to reach the top at 112.5 meters. They looked out over the old city and surrounding countryside. I chose not to go up and instead spent time in one of the many bookstores along the canals. I remember looking up to the very top of the tall tower and seeing a glimpse of my wife's red jacket and thinking, "You are home."

We showed the children where I grew up in a beautiful rural area. I showed them and explained the happenings of the last days of the war. The cellar where my family had sought safety. I showed them the window through which shots had been fired--amazing that no one was hit. It all might have been more than they could grasp. I don't know. I do know that when my wife, at the end of the long day, asked Joshua and Jacob what they had liked most about that day--the market, the church bells, the countryside, the windmill they could climb, the church where their great-grandfather had been preacher--they said, "The stories Baba told."

Later, after they had flown back to New York, I went back to where I grew up to attend the memorial service at the Canadian War Cemetery near Holten. Arriving early in the morning, I parked my car and walked from the small town, crossing the railroad tracks, along fields to the wooded hill where the cemetery lies with the graves of the Canadian soldiers who died while fighting through that part of the Netherlands.

The eerie sound of bagpipes drifted through the trees. Several of the buried soldiers died close to our home during the fierce fighting. Attending the memorial was emotional for me, especially seeing the thousands of veterans who had once more come across the Atlantic to be together with their fallen comrades, and to be together with their Dutch hosts. I talked with several of the veterans and thanked them.

That afternoon I went to a special marker, a white cross near a field where one of many soldiers had been killed. Flowers were lying at the base. As I drove up, I saw a friend from my youth talking with people who live in his old home. He was there with his Canadian wife. A chance meeting if there ever was one. But that was only the beginning.

Later in the early evening, I went to another cemetery close by to visit the graves of my parents and brother. Leaving there, I met somebody who thought he recognized me. We talked. I introduced myself and we realized that indeed we knew each other from over

50 years ago. The old friend told me that the man he had been talking to as I entered the cemetery had been a baby when I was young. And I used to play and help on his parent's farm. Taking the horses to the blacksmith, riding them bareback, and oh, how I enjoyed that.

Another chance meeting occurred when I drove this friend and his wife to their host. It all was close by and all oh, so familiar. But meeting his hostess, who was very busy cooking the evening meal, I learned that she had been friends with my sister.

There were other people whom I met that day who knew me as a child and I remembered them. It nearly was too much. All this in the beautiful countryside where I grew up, which was torn apart by the war–then rebuilt. It all felt so good to be known, to be remembered, to talk about the past, my parents; to enter the church where my father had served, to hear the present organist say, "My father was the organist when your father was the preacher here."

It is true--you can go home!

As I slowly drove away looking out over the rich farmland, I saw green John Deere and blue Ford tractors mowing grass for hay and/or silage. Good weather was expected and they were making hay. I slowly drove the country roads, crossed the big river, and drove on past small villages with their church steeples rising above the treetops. It is an old landscape, changing, but I know where it is still untouched, where it is quiet, where I have footsteps.

Yes, you can go home--only, you can't stay.

May 19, 2004

The Hammock

Several years ago, Anneke, my wife, gave me a hammock. It might have been for Father's Day or perhaps for my birthday. I know it was given to me to be used, but it had stood in the corner of the closet for several years--never, ever used. Whenever I looked into the closet and saw the hammock standing neatly rolled up in the corner, I thought that someday I will take it outside and use it. I owe it to myself! However, soon another year would slip by and next spring the unused hammock still stood rolled up in the closet.

Of course, a hammock belongs to summer garden furniture, bringing visions of a lazy afternoon, of a body slowly swaying from side to side in the moving hammock. A cool drink close by on a garden table where a closed book lies, showing the intention to read. However, holding a book is even too much of an effort on a perfect afternoon, lying in a hammock. Slowly--slowly, ever so slightly swaying from side to side, one feels as if one is hypnotized. Eyes close and a bug landing on your face or arm is waved away without ever waking up. Slowly--slowly, rocking from side to side.

To use a hammock properly, one has to have trees spaced far enough apart to stretch the hammock and to have shade. One does not want to block the slight breeze because the breeze keeps bugs away--at least it helps. The trees need to be sturdy enough so that they do not bend over.

One thing is for sure, a hammock--a real hammock--is never hung within an iron frame as often seen along swimming pools. That type of hammock is in a class of its own. The steel frame does not provide shade, unless an umbrella is part of the fixture. But again, that is not the hammock I am talking about. The garden hammock should be between two trees.

Actually, we have two hammocks. Years ago, my wife made a really good one. It was stretched out between two trees next to the camper we had parked on the farm until the mice took over the camper and my wife said that was enough. So after one more thorough cleaning, the camper was sold to a friend at a bargain price.

Now imagine two unused hammocks. One never used. But it all changed when the twins, Richard and Gabe, came over on Saturday and I had the bright idea to hang both hammocks in the shade of some grown cedars planted many years ago. The twins needed to read and I figured that after an active morning they would like that experience.

Each selected a book to read. After showing them how to sit on the edge of a hammock and swing their legs on board, they both settled in and started reading. I watched them settle in, slowly swinging. Every now and then, I quietly walked up. I noticed that their speech became slow and lazy and it was not long before both had dozed off into a deep sleep with books lying on their stomachs and arms stretched out.

It was a good sight to see--the twins sleeping peacefully while the shade and a slight breeze kept them comfortable. Right then they had no cares in the world. They knew how to relax--let their body and mind go.

Sometime later, they woke up and somewhat groggily started reading again. Just like the butler on the million-dollar show, I brought them a cool drink and inquired if the gentlemen found things to their liking. With a big grin on their faces, they said, "Everything is fine--very fine."

Later in the afternoon they jumped out and picked up on earlier activities while the hammocks were left hanging--empty!

I don't know when it was, that same evening or the next day, that I walked by and said to myself, "Now it is your turn, old man, to stretch out in one of the hammocks. You have been busy, you really owe it to yourself."

I walked over to the hammocks and eased myself into one. I looked up into the treetops which slowly swayed in the breeze. The sun filtered through but did not bother me. There I lay, slowing swaying, relaxing--relaxing. But let me tell you, the more I relaxed, the more tasks came to mind, and tasks which I needed to do. It was like taking the lid off a barrel and looking inside and only seeing what needed to be done.

A hammock is not what it is said to be. A hammock gives you a guilt complex in no time. At least it did me. There were weeds to pull. There was a lawn to mow. I needed to write an article. The more I lay there and relaxed, the more things came to mind which I had to do and should be doing now.

Soon, lying and slowly swaying was no fun anymore and I raised myself up and jumped out. Actually, you fall out of a hammock. I walked away and did not look back to the slowly swaying hammock which I had set in motion when I got out.

The moral of the story, or rather the lesson to be learned, is that you cannot relax in a hammock at home. Because the more you relax, the more you see things which need to be done. It just does not work. What does work is this: take your hammock and go someplace far away from home. Put your hammock up in a place where you don't have any chores to do. Then you can relax.

So, if you see an old blue pickup truck going by with a hammock thrown in, it is me looking for a place to hang out--to relax!

April 6, 1994

Planting the Bur Oak

Someday cattle will lie in the shade. Horses will stand under it, slowly switching their tails on a hot, summer day.

It was only a matter of another year or two and the crown of the young Bur oak would reach the overhead wires. The last couple of years, each spring or fall I had told myself this tree should be dug up and moved. But there seemed to always be something that came in-between. I was too busy, or it rained or snowed, the ground was too wet and before you knew it, the young tree had leafed out again and was sending up more shoots. With its rapid growth spurts, it was just like the parent tree, also a fast grower. The amazing thing is, I am talking about the Bur oak, generally known as a slow-growing tree. But not the one I have.

There are several Bur oaks growing around my home but this one, besides growing under the overhead wires, is special to me. I did not plant this tree. The squirrels must have done it. But I know from which tree the big, burly acorn came. It came from a Bur oak not ten yards away from it, a tree now twenty feet tall which I planted as an acorn. Having planted it as an acorn and cared for it, the parent tree holds very special meaning to me.

I like to grow trees. When I planted the tiny acorn with its two tiny leaves already showing and the tender white root, I placed it near a shed yet to be built. Now years later, the shed is there and the growing tree adds character to that corner of the yard landscape.

I wish there was assurance that no one in the future would cut this tree down. But I have lived long enough and seen enough change to realize that very little is sacred. The tree I had planted as an acorn is too tall to move but the offspring, now ten feet tall, gives me a chance to perpetuate this tree line. It is a little bit like recognizing the bloodlines from a good bull you raised.

But before I dug the tree, I had to know where to plant it. The tree needed to be placed where it would grow to its full potential, away from overhead wires, away from gas lines, away from widening roads. Also, the place should be accessible so that I could drive my truck close to the planting hole as the root ball would be big and heavy. I want to be sure this tree lives, hence the big root ball. The rule is that for every one inch of trunk diameter you should dig ten to twelve inches of ball. This two-inch caliper tree needed a 24-inch ball and that is heavy.

During last week, I anxiously looked at the buds to see if they were breaking. The best time to move a tree is just before the breaking of the buds. Other trees all around the yard were already breaking bud and some were showing color, even full bloom. Well, those foolish trees learned their lesson these last few days with the recent cold snap. Their blossoms froze and they don't look too whippy. But the Bur oak, a smart tree and well adapted to our climate, holds its buds tightly closed until danger of frost is over. It is not fooled by bursts of spring weather followed by the typical cold spells.

So the buds of my young Bur oak held tight and this past Saturday morning I looked for burlap bags, sharpened my spade, and set out to dig. This young oak grew up in the border among other shrubs so I had to wiggle my way around it, quickly wrapping up the lower branches so they would not hit me in the face. Then I drew a circle around the sturdy little trunk, giving it a good 24-inch ball, and started to dig carefully.

Pushing the spade all the way down, I cut the surface roots. Then I turned the soil away, spade by spade, slowly releasing the ball. Once I had gone around the ball, again I pushed the spade down to cut deeper roots and again dug more soil away. With a firm, round ball standing in the hole, I took a short length of number nine wire and laid it behind and against the base of the ball, hooked it up to the bumper hitch of the truck, and moved the truck slowly forward, cutting the taproot clean and neatly. The tree was free to be lifted out.

I carefully wrapped the burlap around the ball and waited. Around coffee time, the boys came home and together we lifted the tree out of the hole and hoisted it onto the truck bed, securing it so the tree would not roll over and break the ball which would damage its fibrous roots.

Filling two cans with water--together with my son-in-law and grandson--we set off for the farm. By now I knew exactly where this one specimen tree was going. We planted it on the farm in an open field with good bottom soil. There its crown can reach all the way to the sky, its branches can reach out in all directions, and its roots can go wide and deep, anchoring this native against storms yet to come. We staked the tree and mulched it with some old alfalfa hay.

Dean and Bud will shake their heads when they see it. Now they have to work around the tree. But I love it! It's a little, big tree, a Bur oak on a five-acre field. Someday, cattle will lie in its shade. Horses will stand under it, slowly switching their tails on a hot, summer day. Birds will nest in its branches. A hawk will perch near the top. I see it all.

As my youngest son said, "It's like in the old country."

July 29, 2009

The Visitor

Her voice did not show any panic or anxiety when she called downstairs for me to come upstairs and help her. "There is a snake here, and I almost stepped on it," she said. I quickly went upstairs and asked where it was. "In your room, behind the door," she said.

When Anneke called, I expected it to be a small garter snake. But lo and behold, when I opened the door, I saw a real snake! A beautiful, smooth animal lying curled up behind the door where it was dark. I could have grabbed it but chose the walking stick a friend gave me years ago. In no way did I want to harm the snake. Nor did I want it to slither further into the house, go behind my bookcases, or disappear in the slight crack along the floorboards and escape under the subfloor.

I gently used the walking stick to guide the snake back to the hallway to the opened door. It saw the light and slithered across the threshold, the concrete, the steps, and slipped in under the large flower box. He or she was outside again and free to roam.

I have no clue how the snake got inside the house or how long it had been inside. I wonder how long that snake had observed us. I can just imagine the critter crawling in bed with us. Or slithering up on the cupboard among the dishes or curling up in a pan, lots of possibilities. I wonder how long it could have lived behind my books. Would he have liked the classics, or perhaps poetry? My guess is he would have rolled up behind Louis L'Amour's book titled *Frontier*. That would be the place to be for a real snake.

Because it was a real snake. And don't accuse me now of telling tales or bragging and stretching the size of the fish caught on a fishing trip--spreading your hands wider and wider each time you tell the story.

As I maneuvered the snake out of my room, back into the hallway, and out the front door, I could observe the snake in all its glory. It was a beautiful animal.

After it was gone, I grabbed a book to see what kind of snake it could have been. It was at least two feet long. The color was a shiny black with yellow stripes. From what I read, I think it was a red-sided garter snake. It fits the description and its overall length as they are most commonly around 22 inches but can grow up to 39 inches. It is found in a wide variety of habitats. I can say "yes" to that one. Habitats include marshes, wet meadows, margins of ponds, woodlands and woodland edges, flood plains, and cultivated fields. It prefers areas with moderately moist vegetation.

Of course, with all the rain we have had, it is paradise here even on the hill. The grass is often wet, easy to slither through, and the big flower border is a humid haven. With the rocks and boulders down the south slope, there are plenty of hiding places to give birth to their young in late summer or fall.

The number of young per litter varies from seven to 80. I'll settle for 25. After all, our house only has four bedrooms but thirteen bookcases, if that is what they like. I looked up what food they prefer--frogs, earthworms, toads, and small mice, as well as other small snakes. In turn, they are eaten by larger snakes and the hawks.

I wished I had picked up the snake to help determine what it was. Continuing to read, I am also considering the lined snake. However, the lined snake is only seven and a half to twelve inches in length and our visitor was much bigger--at least two feet!

What struck me was how beautiful the snake was. As it moved, it had no traction on the wooden floor in my study. Nor did it move fast on the concrete. Compared to snakes I have observed on the farm, it was definitely challenged.

The other thing I noticed was that it was not afraid. Yes, the snake moved away from the gently moving stick I used to direct it to give it back its freedom. But, it did not panic. I wonder how good of a pet such a snake would make.

Family in Norway had little adders, poisonous snakes, living under their house. But we all knew they were living there and that was fine. I remember in Australia, a huge black snake living in the ceiling of the farmhouse. You could hear it slither across the ceiling. No mice lived up there!

I have seen and observed several snakes on the farm. They are big snakes and not the type to select as a pet. The snake which came to visit us was different. It seemed to feel at home. I wonder how long it had been living with us.

August 19, 1998

Deer in the Yard

For a doe to have twins is common, but triplets, I like to think, is special.

The August 1998 issue of Readers Digest has an article entitled "When Nature Comes Too Close" by Anthony Brandt. It shows a drawing of someone barbecuing in the backyard while a fox, a deer, a moose, and a bear look on from the edge of the woods. The article describes the invasion of wildlife into man's backyard. Of course, let us not forget we have moved into their backyard--their habitat.

"To attract deer, plant apple trees," I wrote years ago on the margin of a page in a book on wildlife management. I had picked up the book for 99 cents on sale and every now and then I'll browse through it. The complete title of the book is *Practical Wildlife Management* by Dr. George V. Burger. I have always been able to locate this book instantly among all the books on my shelves. It is never very far from me.

This weekend, on Saturday afternoon while I was relaxing and drinking a cup of coffee, my wife suddenly whispered, "Look--look." Not being one to jump up when people yell "fire," I casually looked out the window in the direction she pointed. I could not see what excited her until I noticed the twitching ears and recognized the camouflaged body for what it was--a fawn eating from the fallen apples under the apple tree next to the sandbox. The fawn blended in with the soft sunlight which filtered through the branches of the cedars.

The fawn was no more than ten yards away and rather casually nibbled on the fallen apples, taking one in its mouth every now and then to crush it. I could hear it through the open window. We watched the fawn move slowly among the trees. Then suddenly a second fawn stepped out from the shrubs along the windbreak border and soon was followed by a third. They too started to nibble on the fallen apples.

All three fawns still showed the remnants of white spots, but they were fading. They were very much at ease, which surprised me. What also surprised me was that our dog, Kansa, did not pick up their scent, nor had any inkling of what had invaded his home territory, making themselves at home so close to the house. After a little while, they walked off, disappearing among the shrubs.

Sunday afternoon, around the same time, they visited us again. The first fawn went immediately to the apple tree and like the day before, started to munch, soon being joined at the feast by the other two. As before, they paid little attention to their surroundings, feeling quite safe.

Then, to my surprise, a large doe stepped out from the cedar trees and joined the fawns. She showed no fear but was much more alert. When I clicked the shutter on my camera, her head immediately turned my way--toward the open window. She looked and stood rigid with her ears out like sails. The fawns continued browsing among the fallen fruit and never paid attention to the doe's alertness.

I did not move but stood silently watching her. Only when she relaxed and slowly moved did I take another photo. Again, the shutter sound made her freeze. Now the doe became somewhat frustrated and started to move away. However, she was not in a hurry but walked the path to the meadow--with the three fawns following her--no hurry. The four of them walked into the fading sunlight, nibbling as they went.

I am not surprised that deer visit our property. I like to think that our four acres are the ideal range, which is 50 percent brushland, 25 percent woodland, and 25 percent non-forested or grassland. The vegetation on our hill is slowly changing, making it more attractive to deer. However, I am surprised that the deer feel so much at home that they step out in the yard and eat from the fallen apples. They must know there is a dog--a dog which loves to give them a chase. Are the deer so confident that they know they can scatter and run?

The fact that our four acres are on the edge of town, adjacent to wooded hills and rangeland, helps explain the abundance of game. They are attracted to certain food, in this case apples, which makes for very interesting observations while drinking a cup of coffee. That the one doe has three fawns I see as the result of good nutrition. The habitat is excellent for deer and food must be abundant.

I am fully aware deer can become a nuisance in the garden. But if you apply a kind of naturalistic landscape approach, little damage is done. They are part of the picture--just like the red fox, the bobcat, and the chatty squirrels.

September 2, 2009

The Murder

Somebody has been murdered on my hill. All I can say is that I found the evidence when I walked the trail and for some reason, left the narrow path and stepped into the tall grasses. There was tall bluestem reaching to my shoulders and Indian grasses, some goldenrod mixed in, with a few yellow sunflowers.

What caught my eyes was the flattened grass and I thought that deer had bedded down close to the trail. But I was wrong. There had been a crime, a murder. The evidence was there. No, there wasn't a body or any remains, only some clothing or what you could call clothing--bird feathers. In this case, turkey feathers scattered around in the bedded grass. It seemed there was a brief struggle and then it was all over.

I stood there looking around me, but the culprit was not there. He was long gone. And the only evidence was--the feathers. I could tell what had been murdered, but not which one of the several turkeys we see wandering around our house. The turkey had walked up the trail and maybe even rested in the tall grasses. Bad move. There is danger in those tall grasses, especially close to the clumps of dogwood and sumac. That is a great hiding place.

I tried to reconstruct the misdeed. Not too difficult to see where the bandit had been lying in wait. I wondered if he was the same one who swaggered around the house at dusk and had started yapping under our open bedroom window a few weeks ago. His yapping woke me up. I was about to call the sheriff when he shut up and slinked off, very typical for his type, moving around undercover in the darkness. But my grandson, Matthew, saw him running around during daylight. He was helping me unload some hay and remarked about some crazy coyote walking the hill.

So, it could have been that coyote or one of the pack. Because I know there are more. There may be a den. And if there is, then there might be pups. If there are pups, maybe a coyote was only shopping for food and the turkey was in the wrong place at the wrong time. Maybe it was not murder.

I have to go back to where it all happened. I must admit there was no waste. None. It was not like when our cat was taken and I found half of it eaten and the other half just lay there. I buried the half I found. Maybe I should not have as the coyote might have come back for the leftovers. Having buried it, I placed a rock over it. That was Pee-po. The cat had a fun life but made the mistake of straying too far from the house. It is dangerous, rough country.

But the turkey was taken all together and only some wing feathers were left to let me know what had happened. I won't call it murder anymore. I like the term was "harvested." Now I know what was taken and I can only guess how it was taken. A jump and a quick snap of big teeth, a silent death with some flapping of the wings. If there were more than one turkey, the others would have run or flown away with heavy wing beats.

It must have been during the day because in the evening I hear the turkeys go to their roosting tree. They fly up, settle down, and chat for a while, telling stories of what happened to them. I won't call it murder anymore. They all must have sat quietly and thought that it could have been any one of them.

I don't think coyotes look at turkeys and say, "Now that is a nice, fat one. That is the one I want." I expect that they are opportunists and grab what is closest--wwhalp! So being a skinny turkey isn't an advantage. However, if I weighed the turkeys of the flock roaming around our house, I expect they all have good weight as they hang around the bird feeders and pick at what spills over. These turkeys don't look skinny. Corn and sunflower seed is their daily menu, as well as the insects on the hill found among the tall grasses.

But I must be fair. Presently I am suspecting that the coyote is the murderer. But I have no proof. If I find coyote scat with turkey feathers, I have proof. But we have had bobcats prowling the hill and yard. The last time I saw one, it was sitting in front of the border grooming him or herself. Another possibility is the fox. I watched a fox play in the yard last winter. I have not seen him since he pulled a stack of wood over, looking for a packrat hidden there.

So I am back to the coyote, the trickster. I know there are people who passionately hate the coyote. They accuse him of all sorts of misdeeds like killing lambs and calves. I also know that they clean up a lot of cadavers. I have proof of that. My horse that was struck and killed by lightning was gone in two weeks, cleaned and bones scattered. One of these days I'll pick up a bone and hang it up somewhere as part of a good memory of Flash.

I might walk up the hill and pick up the few feathers which are lying in the flattened grass. I am sure the coyotes don't care. There is absolutely no meat on them. Just feathers, beautiful turkey feathers no longer needed for flying or bragging when displayed by a tom turkey.

May 26, 2010

Growing Cut Flowers

There should have been small, tiny plants visible now, although difficult to distinguish from little weeds. The only thing that helped is that we planted the seeds in a straight row and marked each end with a white tag. There are five such rows. The seeds we carefully planted are annuals. Each seed holds a promise. It was my idea as a Mother's Day gift instead of a big bouquet of flowers in a vase.

I had the space, good soil in the pony corral with plenty of horse manure to add as organic material. It was all dug up as my grandson, Derek, started digging there months ago when he considered growing vegetables. I had shown him how to dig scientifically. That is, to dig a straight, deep trench. Next, scrape in manure and other organic material matter, then dig the next trench, placing that soil on top of the green stuff. It works beautifully when done right and gives clean garden soil. However, the vegetables did not get planted so I made other plans.

First though, we had to reclaim the area as the weeds had grown hip high. But with a little energy, the patch was cleared and the space reorganized. Then it was time to sow. Derek helped, learning as he went.

The five rows are well separated so that one can walk through the rows and cut the flowers. It is to be a cutting garden with old fashioned flowers. I don't even know what all we planted. I do know there are two rows of sunflowers and then some different packages of seeds to which I added some seeds from the birdseed can. Now that should add a touch of wildness. Being a cutting garden, one can be kind of ruthless and cut whatever one wants.

Before planting the other annuals, Derek and I mixed the seeds with some good soil in a bucket, stirring and stirring until we could not see many of the seeds anymore and they were thoroughly mixed in. Some seemed as small as alfalfa seeds, others were larger. Knowing that roly-polys are always a problem and a challenge, I decided to mix some marigold seeds in too. I have no clue where the marigolds will come up in the rows but I surely hope in regular intervals so they can do their job and keep the pests away.

Like any sowing--be it wheat, corn, or beans--one has a tendency to be very happy and satisfied once the job is completed and you begin to look forward to a good crop as a reward. As opposed to a good hay crop or plenty of bushels of wheat, corn, or beans per acre, my reward will be rows of cut flowers and watching Anneke go out and harvest.

Watching the "going out" is literal as the coral is at the end of the winter paddock, next to the old stone wall. It is a well-protected site with trees on the west side and good sunshine during the day. With a slight slope to the south, it is perfect for growing flowers. It holds promise.

I will admit that my garden strolls have already taken me there several times. I'm like a farmer driving a detour along a recently planted cornfield, knowing full well that no shoots can be up yet. Still, the promise is there and the expectation. Whenever I look across fields and see a bare row among solid rows of corn or wheat caused by a blocked seed pipe or a rock pushed along, I cringe even though they are not my crops. I hope I do not have any gaps in my rows. I expect solid rows of flowers--mixed flowers--to come up. It is still too early. "Patience," I tell myself, of which I have little.

As I go on my walk to the flower garden, I pause and give my greetings to Henk. He is my alter ego and has a job to do--keeping the deer and turkeys away from my flower nursery. Henk was cloned using a six-foot two-by-four. Placing the two-by-four in front of me, I marked the height for my head, shoulders, and hips. I nailed cross timbers for my shoulders and then gave Henk some of my old clothes. I considered underwear but thought better of that. He has an old, holey shirt, old jeans, a scarf, and one of my old hats. To take the shirt in and hold the jeans up, I braided six red strings of bailing twine into a strong belt.

I placed him on the fence near the rows of flower seeds. Henk looks kind of scary--like a good scarecrow should. But what is really scary is that he looks like me, six feet tall. But my clothes on him make him me. The sloppy hat gives him that silly look so typical of scarecrows. I have called him Henk after an acquaintance of mine. So now when I work there, we can hold conversations. A polite, "How are you, mate?" generally starts us off on the right foot. I thought of placing an empty can of beer in his hand. He must get thirsty on a hot day but I don't want him to look too much like a hobo. After all, he looks like me.

"Now Henk, remember we have to grow flowers. You do the looking and the scaring and I'll do the work, mate. And when my wife comes down and starts cutting flowers, just hush up. No need to tell her anything. She knows what she is doing and remember, this is her project. We grow these for her. It is her cutting garden.

Yes, I hear you. Maybe Derek and I should never have started this project but hey, you are sitting on the fence now so keep the deer away. I want to see those flower seeds grow."

September 7, 2011

Going Places

It was a father and daughter vacation and they bought two airplane tickets to Ireland, their destination. However, they went with the three of them. Apparently, the airline did not mind as they were never stopped. Oh yes, they looked at their luggage, shoes, etc. but they had no problem boarding the plane where they took their assigned seats, two seats. As their plane taxied to the runway and started taking off, they sat in their seats thinking, "Here we go, just the three of us." We are off to Ireland for our dream vacation.

Plans had been made for a long time, but not all the details as spontaneity would be lost. As a matter of fact, they had been to Ireland before. Ireland is a beautiful country with a long, rich, and sad history. Many Irish came to the United States. They worked hard, very hard, and were successful and raised their families. But deep in their hearts, they remembered Ireland, the old country.

There is no Irish blood in my granddaughter. On her dad's side, they are Swedes and her mom's side, Dutch. An interesting combination but does not really explain her fascination with Ireland. I think what intrigues my granddaughter is the mystique of Ireland expressed through its ballads and songs, which reflect the Irish landscape, its people, and history--even its losses.

On arrival, after their transatlantic flight through the short night, they picked up their rented car and went their way. But first, they installed their third passenger, they made him comfortable.

The next two weeks, the three of them crisscrossed Ireland, looking at the old and sacred places and enjoying the Irish coffee whenever they pulled over. I don't know what conversation they had with their third passenger. Whatever they told him, I am sure he did not answer. I don't even know if they took him into the pubs. Probably, "No dogs allowed." However, I don't think even that message would have stopped them with Wheeler being so dear to them. Because it was Wheeler who they took along, their intensely sweet Labrador who died a few months before and was buried on the farm near the gate with the sign "Joni's Fence." Yes, the spirit of the dog went with them to Ireland. And not only the spirit but they took the dog's collar and hung it in their rented car. I can just see it.

I too had a very special bond with Wheeler and at times, I have a feeling he is still here. I have heard him walking the wooden floor and sigh when lying down. When he was young, it used to be a sigh of great joy and satisfaction. A sigh only a contented dog can let go when plopping down. Later, the sigh had some pain with it as old age, arthritis, and weakness set in. But for a long time, he would not give up.

Throughout his rich dog life, I have "dog sat" him. He enjoyed coming to our house and when young, coming with me to the farm. There, like all dogs, he ran, searched, and dug furiously for pack rats. Only once in a while, he caught one. But he never gave up. His paws would rattle the sticks, his pitched breathing turning into an excited yelp. I would encourage him, "Go for it, Wheeler. Get them." After a while, he would go to the pond and swim. The cold water did not bother him when he was young.

Then old age took over, he had lived a full life and given such joy. He had taken care of the house and especially the children. As they grew up, each of them had a soft spot for him and he for them. At bedtime, he would find his place. For a long time, his favorite spot was with Joni, under her bed. So it was no wonder that Joni decided Wheeler had to come along on this trip with them. It was his ghost that made the trip with them.

After two weeks of exploring Ireland, they returned the car and prepared for home, ready to come back. They removed the dog collar and put it back in their suitcase and slipped through customs again with no one noticing there were three of them.

Flying back, they must have been quiet, completely satisfied with the time the three of them had enjoyed together. Here, in the U.S.A., customs let them through. No questions asked. "What about the ghost of this dog?" No. No one noticed.

Back in Kansas City at the airport, the dog jumped easily into the family car. He knew his place. The old place where he always had traveled. He was a good traveling dog. Always had been a good dog, a trusted friend--Wheeler.

Back home, they unpacked their suitcases. They hung Wheeler's collar back on the hook. When Wheeler died several months before their trip to Ireland, he was buried on the farm near Joni's gate. Father and daughter did the job. It was hard digging. They chose the spot because it held special meaning to Joni. The gate, an entrance wire gate onto the range, is part of the internal fence Joni and I built several years ago.

Next to the gate is a growing cedar with a deep green color. The soil is reddish, indicating iron. It is a good place for Wheeler to rest. After all, it was a long trip for a dead dog. He was going places, but now he is back home. In Kansas soil!

October 5, 2010

The Rug

Saturday afternoon the thunderstorm came roaring in. When I saw the purple-green clouds stack up, I went into the house to grab the camera to take a picture. The colors were awesome.

However, back inside, the storm was already above me and I could no longer take the picture I had wanted to take. Thunder roared and lightning flashed. Later I learned that the football game was delayed. Smart move. The rain and wind gusts blew around the house and then as quickly as it started, it was over. The blue sky came back as did the sunshine and I heard the crowd cheer again. The game was back on.

As I looked at the still dripping and glistening leaves, I noticed that a rug had blown onto the front lawn and was now lying in the middle of the open grassy field. The trouble was, we did not have a gray-yellow rug lying about for the wind to pick up.

As my wife was standing beside me, I said to her, "See that rug? Keep your eye on it as I go outside." When I opened the front door, she started laughing. The rug had come to life. It stood up, shook itself, and disappeared through the flowers. All I saw was the coyote's tail. Because that is what it was, a coyote sunning itself. What I thought was a rumpled piece of rug was a coyote lying on its back with belly up to the sun and the four legs thrown about as if he or she had no care in the world. And I am sure he/she had no care, it was home.

Later, my wife saw two young coyotes. So I expect they found a suitable den among the boulders and rocks down the bank. As long as there is game to catch, I expect they are happy to stay. They have woken me up yipping at the full moon. I wonder if someone is paying them to sit under our open bedroom window. It seems that is where they like to sing.

We have not had coyotes for quite some time and it has caused some imbalance in nature. I have taken care of raccoons, possums, and even squirrels. Their numbers are down. But there are plenty of rabbits. As I pulled weeds in the flower border earlier this summer, baby rabbits would hop in front of my searching and weeding hands. They were hiding among the foliage and did little damage to the plants. At dusk, the families would gather on the lawn and do their nibbling on grass, dandelions, clover, and apple peelings we tossed out.

But now, there soon may be a rabbit scarcity. I expect turkeys will be next on the menu with a juicy packrat thrown in for a tasty tidbit. I would pay the coyotes to go after the packrats! Really, I would! A few years ago, we had a hungry coyote come up close to the house to eat the apple peelings, a very health-conscious beast.

I enjoy watching the antics of the coyote and he or she--the whole family--is welcome. We have no cat. But I have worried about my neighbor's cat. "Look out, honey. There is danger here." Having looked into her fierce yellow slit eyes, she might have a chance. However, if I were to bet, I would say someday she will be caught for dinner. But don't blame me. I warned you.

I am sure the coyotes will watch me from among the trees where they feel safe. I hope they get used to me, realize I mean no harm, and show more of themselves. I might have to buy some chickens and make a coop. I wonder how high they will jump or climb a tree to get what they want. It surely adds interest to a yard. This coyote family is a long way from the open space of the farm where coyotes cleaned up my dead horse and scattered the bones.

In comparison to the coyotes on the farm, my front lawn coyotes are urbanized so much that when I hear the sirens of police cars, the coyotes join in and answer with their chorus. It is quite amusing to me. I can just visualize the coyotes sitting outside their den responding or stopping during their early morning prowl, lifting their heads, and joining in the wailing sound. It is fair warning to the rabbits and turkeys. Unless the yipping and howling sparks the curiosity in them.

I have counted the wild turkeys as well as our rabbits and our neighbor's cats. We'll see how long the supply lasts. I am sure nature is balancing here. It happens even in the urban environment.

Many ranchers look at the coyote as a varmint. Pickup trucks hold rifles and ammo often rolls through the cab or along the dashboard. I like the coyote, the trickster, and whenever I hear or see him on the farm, I give him his place. He belongs.

Mary Austin, a naturalist, once wrote, "No man has ever really entered into the heart of a country until he has adopted or made up myths about its familiar objects." I like to think I have.

December 19, 2007

Candlelight

Monday morning there was a clear blue sky, frozen ground, and ice still hanging on the trees. Even with the bright sun shining there was as yet no drip--drip--drip of melting ice. It was still as cold as it was in the early morning.

If you had to feed, I hope you had a closed cab on your tractor and that the engine kicked over when you turned the key. Sometimes they don't start, at least not immediately, sometimes not at all and then it might take some tinkering to get the old, trusted tractor going. But once it goes and responds to the throttle, you can demand power to lift the large bale in the front or in the back.

If you are alone, you'll have to climb down and open the gate to let yourself through while cattle are pushing impatiently on the other side. The same ones are always in the front line--eyes rolling. They are hungry and cold. The good hay will warm them up from the inside out as will the grain or whatever else you feed them.

Anyway, when all the chores are done and things have settled down, you could run into town and warm yourself in the coffee shop. There are already several pickup trucks parked there. A few carry a big, round bale besides chains and shovels covered with snow and ice. Those are just for an emergency. Driving is still good with the ground frozen. But once the ground starts to thaw out, that is when it becomes more interesting--or when driving on hard ice, slick ice.

When I talked on the phone to my son in Canada yesterday, he was pulling an eight-horse trailer with his big Ford truck and he suddenly broke off. When he called back later, he said, "Sorry, I was sliding." Oh, winter is fun, especially when you care for livestock.

This Monday morning, when I stepped outside I thought I still heard a generator or two working in the neighborhood. They have been plugging along ever since we had the ice storm come through Manhattan. We were lucky. Our power came back on after two and a half days. It does not take much to appreciate that. With no electricity, life becomes simple. There are lots of things you cannot do. Where electric power is required on farms, it becomes hard labor unless you have generators.

With plenty of firewood, I lit up the fireplace and started warming the chimney and the living room. We are lucky that it is a good fireplace, throwing a lot of heat into the room. We enjoy snuggling up to it in the evening. I stuck my socked feet in front of the screen. After a while when my feet got too hot, I pulled back, enjoying the now hot socks.

It is a simple life without electricity. I sensed how simple when I had gone outside to bring in a load of dry firewood in the wheelbarrow. It was getting dark and I had to push snow and ice loaded branches away from the wood stack. I reminded myself to cut those low hanging branches off when the snow and ice melted or move the woodpile.

I pushed the loaded wheelbarrow into the now dark garage, stomped the snow off my boots, shook my warm cap, and stepped inside. It was at that moment that I realized that the Christmas season is here, that Christmas is getting closer.

When I stepped inside the warm room, warmed by the wood fire, the soft light in the room was--candlelight. Anneke, my wife, had gone around and lit the candles which stood on the furniture or in the large candle holders on the ground. The soft yellow glow with moving flames threw a beautiful light in the room and was reflected in the large windows. It was dark outside, winter with snow and ice. But I still see the dancing candle lights reflecting in the wood of the furniture. It was very, very quiet stepping inside. That is when I thought, "This is Christmas." We sat with burning candles that evening.

The next day when the power was restored in the late afternoon, the furnace kicked on and it did not take long for the house to heat up and be comfortable again. In the evening when it got dark, I flipped the switches and electric lights lit up the rooms. And out in the workshop, I had power. It probably won't be long and I'll take it for granted again. I expect it. I don't know better. But yes, I know better because I can still feel the awe I held for the candlelit room.

May your electricity stay on so that your work is easier and you can see where you are going in the dark. But I hope and wish for you that one moment of reflection when candlelight takes over and the light becomes soft and reflects.

Maybe you should at some time throw the main switch and darken the house. Light up your home and live with candles--the soft, gentle light.

January 20, 2021

Landscapes--People

Sometimes people's character reminds me of our landscapes. I am not thinking of only Kansas landscapes. No, I am thinking of our country, the U.S.A. And I am thinking of all our people.

Of course, it is a crazy or silly thought. But think about it.

Every now and then you meet a rugged, strong personality and then I think of our mountain landscape. Such a person can dress the part, but not necessarily so. Just like our mountains, he or she can be a quiet person with deep, steady thoughts like deep valleys or sudden canyons. The valleys are wider and smoother, productive. The canyons, more sudden. Maybe more abrasive and more difficult to penetrate--deeper thoughts. Might be more explosive, more solitary.

Of course, our landscapes have been partly created by our climates. Storms and rushing rains, cold and heat have had influences. So have our characters been influenced.

Like any landscape, vegetation adds character. I am not talking beards or long hair. But again, it is character. We have in the far northwest the deep, dark forests. When the rain blows in from the sea, the trees are misty and moisture is all about making things grow--silent and somber and dark. It can be beautiful and mysterious.

In the mountains, on top, you can look far. Sometimes farther than you can see. The same with people.

Some landscapes are extremely productive. They are landscapes producing wheat, corn, and other crops. They are like steady people doing their task. There can be storms roaring across those landscapes. And the storms can leave scars. But generally, the wide-open landscapes stay, producing crops year after year. They are steady people. They know what to do and what is expected from them. And they do it. If there is adversity, they know how to handle it and help each other.

It is an old landscape with a long, old history and it has seen changes. The changes have formed the characters of people. They are quiet people making communities. They are surely not plain--but steady. And it is the steadiness that gives them strong characters.

We have landscapes with volcano action. It can be explosive. There always seems to be tension. One never knows. Sometimes it comes to a boiling point and characters explode. Sometimes with disastrous results. It is not easy to be such a landscape. You don't know what can happen. It is unpredictable. Yet, it can be a beautiful landscape. The process of healing for such a landscape can take time. In the process, we learn. But things can be destroyed. It is hard. It takes a lot of time to heal.

There are landscapes that are hard to enter. I am thinking of swamps. They are dangerous landscapes and only the people who know them can or should enter. There are characters like that. They may be hard to deal with. Maybe they are misunderstood. Maybe we don't try hard enough to understand them. When we alter the landscape and drain the swamp, we may be destroying something.

There are people like deserts who bloom after a rain squall. All of a sudden, dry, desolate land turns colorful. Those are people who are encouraged when loved, and respond.

The thing I have learned by looking at many landscapes is that all landscapes are connected and that they slowly or more rapidly flow from one into the other. That can be very interesting. We know and recognize the edge effect in landscapes, and we know that edges are the most diverse landscape and often the richest. Maybe that is where the artists are. They may be painters or poets. They may be very creative people. Maybe autistic.

You can think of where the Plains meet the mountains or sea, ocean meets the land. Ask yourself, where does land meet swamp? Think about the bayous near the mouth of our big rivers.

And that brings up another thought. How many landscapes have we destroyed? Sometimes we have destroyed out of ignorance and sometimes willingly. It is the same with characters of people. It might have taken generations. As it will take maybe generations to heal.

If all this sounds kind of philosophical or weird, just think about it as you do your daily chores or work. Look around you and ask yourself where you feel happy or where you long for. It may surprise you as you look at yourself. It may help you understand yourself.

But just like land and landscapes, we are all attached. We flow from one into the other. Yes, I am aware there are islands. As there are people who are themselves. As an island, they have their own character. They can be different. They are no better or worse. They are an island. They are unique.

Those people may be hesitant to come ashore. But remember, we are still connected just as the core of the land is still connected deep under the water. Sometimes not so deep. But we are connected.

People are like landscapes. I sometimes wonder what my landscape is. I am still trying to find out.

Camping Trip

Thanks goes to my granddaughter, Miriah van der Hoeven-Essman, who allowed me to use her essay, "Camping Trip," as the end vignette for this book. It was a Christmas 2020 gift under the Christmas tree; a beautiful cedar cut on the farm in the fence row.

Should I start at the beginning? But which beginning? Should I tell of the great inland sea that left fossils scattered across the fields and creeks? Should I honor the indigenous people who tended this land first? Or maybe I should start with something closer, the first time I camped on this land over thirty years ago. Maybe I should start with the conversation shared with Russ on a familiar green bench that forever joined our journeys. Or perhaps, the first time I brought each of our children to these windswept hills to visit the farm? Each field and bend in the creek has a hundred stories from my life alone. Each beginning connects to the next infinitely. On the farm, we measure growth like the slow majestic history of a Bur oak. Each beautiful giant, once contained entirely in the smallest acorn.

Friday, November 13, 2020. For some, this combination of days and dates may seem ominous. But as our trusted brown truck headed west, my heart was happy. On the drive, we counted 62 red-tailed hawks, a game my mom started when I was young. There is something of a benediction to seeing them soar or perch, and a thrill to tallying each sighting. I've never guessed the exact number, yet something made me say "62" as we went around the truck cabin.

Bea asked, "What's the prize for being right?"

"The satisfaction of guessing correctly," I replied. And there is a deep and precious satisfaction to have not only been close but in somehow, picking the exact number we would see. It feels almost like the last ingredient needed for a secret magic spell.

Not to be outdone by their fiercer brethren, two flocks of turkey and a covey of quails ignored us as we traveled down the gravel of

McIntyre Creek Road. We arrived at the North entrance of the farm at about 3:30.

It is cool in the shade, almost cold, the promise of night's chill is in the air. The prairie has softened to the muted ochre, pinks, and purples of fall. I remember Oma asking me once, about this time of year, what colors I saw. That query changed my perspective forever. Where I once saw the bleached remains of summer's saturated colors, I now appreciate the subtle beauty of late fall's dried grasses. Bright orange bittersweet hangs in fluorescent swags and tangles, brightening the bare branches of trees and shrubs. Dusky green cedars seem richer against clear Kansas sky when they aren't compared to brighter deciduous leaves.

The high blanket of wispy cirrus clouds in that same sky, promise a beautiful sunset. November's late afternoon light gilds the grass, turning each dry stalk and seed pod into a treasure fit for Midas. That same light is a reminder that setting up camp is our first priority. We collectively decide the best place for our tent and then divide responsibilities to prepare for the nightfall. Everett and I gather dry wood while Russ and Bea assemble the four-person tent. With shelter secured, Russ and Everett start the fire and I organize the camp kitchen. On a simple folding table, I line up a Coleman cookstove, a clear tote with utensils and tools, and underneath, the cooler.

The sun dips even lower and so does the temperature. We'd prepared mentally for the chilly evening, but as reality begins to ice our bones, we are grateful for the multiple fires to tend and gather around. To ward against the foul mood that so often accompanies cold toes and fingers, I decide to make a thermos of tea. Thankfully, the water boils just as the first gauzy strips of pink and purple streak across the sky. Everyone cradles their mug and sips the sweet, spicy chai. The sunset is as beautiful as the skies promised. I watch in silence as the sun sinks below the horizon and for a moment the colors are deeper and richer than I could have imagined.

I reheat potato soup. It's hearty and full of chunks of potato, carrot, and ham. I'd made it the night before with the intention of having something hot and filling and easy our first night at the farm. We gather around the fire, some in camp chairs and some standing, the rich soup warming us from the inside out. Soon we begin the ancient fireside ballet, warm front, pivot, turn, warm back, pivot, turn, repeat.

The blue of the sky is deepening from pale to indigo and soon we spot the first star.

I whisper, "Make a wish."

Bea murmurs, "I wish it wasn't so cold." We gently tease her that now she'll never warm up because she made her wish out loud.

We sit around the fire. Together. More together than we are at home. There aren't distractions. There are only the sounds of the prairie. The crackle of the fire. We listen. To the world, to each other. We wonder about the other fires, the history of campfires, and the people who sat together around them.

When I next look up, the stars have spread, they twinkle and shine against the velvet expanse. Here away from towns and the light that spills from windows and streets, the stars are glorious. Every time I see them against the vastness of the sky, it's like the first time. I feel both finite and infinite.

In the midst of my weighty reverie, Bea is cold. Too cold she says, and she wants to retreat to the warmth of the tent and the sleeping bags. Russ coaxes her to persevere just a while longer. "Let's hike to the top of the hill. Maybe we'll be able to see Manhattan." She'll do anything if he asks, so we walk out of our hollow onto the field. Through the dark, we climb. My eyes make out the familiar silhouettes of Bea, and Ev, and Russ against the varied darkness of land and sky.

At the top, the wind is sharper. It cuts through our coats and makes our noses numb. We crowd together and stare up. It seems like the entire galaxy is visible. We look for familiar constellations, and realize we must reorient with the seasonal rotation of the planet. Finally, along the horizon, we see the big dipper. It's another example of the power of shifting perspectives and I store it deep in my heart.

At 9 o'clock as promised, we shiver as we crawl into the tent. Snuggling deep into the blankets, our bodies settle and we begin to warm as we listen to the noises of the night. The barking of coyotes and rustling grasses become a lullaby as we drift to sleep.

There is something restorative about sleeping outside. Even if the ground is more unforgiving at this stage of life, waking at a campsite is always a special excitement. Russ and I are the first to emerge into the pink morning light. Stretching out the soreness, we grin at each other, remembering the years we feel in our bones and muscles. We welcome the day tending the fires, encouraging flames from the almost cold ashes. I start water to boil, savoring the anticipation of the coffee to come.

We make our breakfast preparations in quiet companionship, enjoying the stillness of the woods around us. In the stillness, we hear the faint sound of the grasses shifting and we look across the creek. There, regarding us placidly, is a magnificent eight-point buck. Watching him as he watches us, there is a sense of something special about this time, a reminder to be present, to be mindful, a moment of gratitude. He peers curiously as we prepare food and then fades quietly into the hillside. I've never seen a deer on the farm. Antler sheds once, deer beds and rubs on trees the silent consistent evidence of the presence, but never a deer.

Before bedding down the fire the night before, we had tucked foil-wrapped potatoes under the coals. While we slept, they cooked. Russ cut them into pieces and added them to the sausage browning in the large cast-iron skillet. Maybe smelling the savory food, Bea wakes up and peeks out from the warmth of the tent. We scramble eggs into the potatoes and sausage and breakfast is ready. Everett wakes up and we start the morning with bellies and hearts full.

Another cup of coffee and tidying up the "kitchen," we all pull on our jeans and boots and set off for a farm hike. The creek is our compass as we meander through fields past the old sycamore. A large hawk circles lazily over the silvery branches. We pause to admire the exposed limestone where the road has been repaired. As we walk on, I idly wonder what the moss-covered rocks have born silent witness to, how many feet have pushed them farther into the earth?

We head toward the creek, crossing near the McIntyre Creek entrance. "Left, or right?" we ask. Unexpectedly, Ev and Bea concur that we should follow the creek bed south. It's dry this time of year, and our feet crunch through layers of dry leaves covering the bleached limestone. We cut up and out of the creek and across another field. As we cut through a wooded gully, we hear a gentle rustle. The beautiful buck from earlier in the morning is staring at us, no more than 25 feet away. He snorts his annoyance at our presence, we've obviously disturbed his daily routine. He wanders away, and we all agree we've never been closer to a wild animal.

This time of year, the smells of the farm are distinct. Earthy and rich in the damp underbrush, dry and crisp in the fields. These aromas connect to memories that span my life. The soft, rich smell of the decaying leaves blanketing the creek bed gives way to the dry, sweet smell of the prairie grasses blowing constantly in the Kansas wind, the delicate, fresh smell of Osage oranges; each of these scents is as dear to me as a favorite blanket or sweater.

We decide to hike to Everett's gate. However, we misjudge our location and it's a bit too far for one member of our hiking party. No one gets left behind, so we turn back and head toward the big pond. We angle down the rock-studded hillside, careful not to twist an ankle, and a giant blue heron takes wing soaring over the rippling water.

On we walk, taking a leisurely pace. We pause briefly to admire the work of a very industrious beaver near the small pond. Past Jonas's fence and then back to camp near the pole barn. We have worked up an appetite so we make quick sandwiches for lunch.

After eating, an insistent breeze tousles the naked branches contrasting with the restfulness that has settled with our full stomachs. Everett sits beside me stripping a stick with an ax. Bea practices her whittling, perched on the stainless-steel truck box like a woodland sprite. I can hear the soft growl of the chainsaw as Russ cleans up the downed branches on the perimeter of the field.

I walk across to Russ, and he smiles as he sees me coming. He has an overgrown cedar in his sights. It has grown brawny and is starting to block the entrance to the next field. As he downs the rowdy, prickly branches, the unmistakable smell of cedar fills the air. I help him haul branches to the edge of the field, making a new haven for animals this winter. As it grows, I think of all the safe brush piles Baba has made over the years. It's no surprise that there is lots of wildlife on the farm. They know they are welcome and protected here.

We walk back to camp together. Bea is still sitting in the truck bed. She's carved her initials in a stick. Everett's boots are posted by the zipped tent flap, a silent indication he has taken our advice and is resting. Russ fixes himself a coffee and Bea makes a midday s'more. I resume reading the book I'd brought. The soft exhale of our breath is the only sound besides the wind through the grass.

Then, Bea whispers, "Deer!" We raise our eyes in unison, not moving a muscle. Just across the clearing is a buck; much younger than the one from earlier in the day. His juvenile rack is asymmetric and his curiosity leads him closer to us. We sit in awed stillness and he wanders, nibbling plants, around the edge of camp and then off toward the pole barn. When his white tail is finally out of sight, we turn to each other wide-eyed. I don't think that he'd ever seen a person. He wasn't scared at all.

Eventually, the spell breaks and Russ gets up. There is always another thing to do. He's decided to make a wood stack for next time and begins clearing a stone ledge. Together we stack dry logs neatly. The orderly rows promise a return and a quick start for the next campfire. We always try to leave the farm better than when we first arrived. This time that means neatly stacked wood and clearing brush from the stone walls.

A familiar sadness settles as we begin preparing to return home to Lawrence. Maybe wistfulness is more accurate. There is a freedom that exists under the blue dome of prairie sky. Going back to town means giving up the ease and closeness fostered by exploration and the generosity of nature.

Like rewinding a movie, we undo the tent, pack up the kitchen, and re-roll the sleeping bags. The truck is packed and we survey our wild kingdom a final time. Like the flattened grass of a deer bed, the tent has left a soft outline. The ashes in the fire pit smell acrid and smoky. Though no one says it, I know we wish we were staying in this protected hollow where the world feels far away.

One final promise, Russ slides over and Everett sits behind the wheel of the big, brown Chevy. He talks out loud, maybe to himself, maybe to us as he stays in the tracks of all the trucks that drove across these fields before him. Together, he and Russ get out and open and then re-secure and shut the gate. A final turkey feather waits in the middle of the road, a blessing to carry with us till next time.

Everett drives us down the gravel road. The speed is variable but he is steady. We wave good-bye as we pass the McIntyre Creek entrance. The golden light illuminates the fields again and the sky begins to flush as the sun sets. We take a left turn for our last stop, Everett's gate. The brush has grown around it and the sign is camouflaged by the sticks and grass. We take turns cutting back the prickly saplings until we've made a clearing.

Over eleven years ago, Baba and Russ built this fence. Everett was barely taller than the native grasses the first time I drove him out to see his gate. Now he towers over the sturdy posts. Sometimes, in the right light, I see the child I've known and the man he is becoming.

I think this version of us is best. This open sky, open heart, strong, and curious version. Here we are quicker to help, to forgive, to give graciously. How can we not? Against the perspective of endless sky and boundless range, it seems impossible to be petty or mean or tight. The expansiveness and generosity of this place connect us to each other and to the love that this particular corner of Kansas brings to so many in our family. Time shared on these fields and in these woods is a chapter in a much bigger book, the story of each of us and all of us.

Thank you.

Western Kansas . . . crops need water.

Afterword

This book has been a long time in the making. Guido, my son, gave it a start. The picture on the cover came through his initiative. Talina, my daughter, did the final selection of vignettes out of more than two thousand possibilities. Then the book lay idle. My friend, Bill Lamont, kept reminding me of the open spot on his bookshelf next to the book *Lonesome Dove.*

When I came across Greg LeValley, Customer Service Specialist in University Printing at Kansas State University, I picked up the threads. He set me on the path to meet Joe Reid. Joe was the catalyst that pushed me forward.

Then when I found my editor, Annette McHaney, it all fell in place. I was no longer a "blind man on a moonless night in the dark woods." Annette has steered this project to completion and I thank her. But remember, I have the final read--any error or mistake is mine and mine alone. The editor is like a captain on a ship. The captain did not build the ship. However, he or she sails the ship with passengers or goods of words safely into the harbor where the ship is unloaded--resulting in this book!

Dan Donnert, the photographer for Kansas State Research and Extension News Media Services, expertly retouched the photo on the back jacket of Flash and me. The photo had many pinholes having been posted near the telephone for years. It was taken by a photographer for the student-run paper, the Kansas State Collegian.

Thank you goes to Sjoert-broer for checking the correct spelling of the Drentsche Aa. It is a small historic Dutch stream in northeast Netherlands.

Now, for all you readers, love and take good care of our state--Kansas! It is very special--it is home!

Kansas sky.

Made in United States
Troutdale, OR
11/03/2026

56844901R00199